Farm Animals
Your Guide to Raising Livestock

by
Jeanie Peck-Whiting

First Edition
Fox Mtn Publishing, Tonasket, Washington

Farm Animals

Your Guide to Raising Livestock

Farm Animals
Your Guide to Raising Livestock

Jeanie Peck-Whiting

Fox Mtn Publishing
PO Box 1516, Tonasket, WA 98855, USA
Email: info@foxmtnpublishing.com
http://www.foxmtnpublishing.com

- Cover Design by Robert Howard
- Edited by Sheri Vejrostek

Publisher's Cataloging-in-Publication
(Provided by Quality Books, Inc.)

Peck-Whiting, Jeanie.
    Farm animals : your guide to raising livestock /
    Jeanie Peck-Whiting. -- 1st ed
    p. cm.
    Includes Index.
    ISBN #0-9716174-0-6

    1. Livestock—Handbooks, manuals, etc.   I. Title.
SF65.2.P43 2002

Acknowledgment

Above all I would like to thank Ron for supporting me during my writing days. Shana, Amorette and now Kyra-Jean were *forever* helping me; thank you kids!

A special thanks to Bill and Sheri Vejrostek. I can't thank them specifically because they were an endless source of information and help. Thank you to all of Bill and Sheri's animals; the pigs and cows who have given me many stories!

Although she doesn't realize it, my gratitude goes out to my #1 sow, Rosie. She has been my inspiration to write the book *Pigs; and other stories* and now *Farm Animals*. I love you, Rosie!

A special thank you to all the folks who have helped me with stories for this book. I can't name you all, but to name a few, I thank Shelly, Don and Jack at Tonasket Feed, Nick Baker & family, Galen & Shannon Garoutte, Cody Ames, D.V.M from Tonasket Veterinary Service, Dan & Debbie at Maverick's Bar and Grill in Tonasket, Ed Maleng, Terry Dean, Sharon Stepp, Gina Koenig, Don and Wendy Mahlendorf, Melissa Maple, Esther Krohn, Doreen Morris of Okanogan has been an excellent help, Jacqueline Walimaki, D.V.M. of Oroville Ark Animal Clinic, Greg Short, Araia of Wauconda, and Meg Lange at Tonasket Library who was always there to help me find the books necessary to produce this one! I know I missed a zillion folks whom have helped me in one way or another, I thank you all.

I give my heartfelt gratitude to my Uncle Don and Auntie Pam Ohman of Burbank, Washington for always believing in me. You both mean the world to me!

A Word from the Author

Wanting to share all of my experiences and knowledge about pigs, goats, and cows and other farming animals and activities, here I am.

I am not trying to tell my readers how to raise their own animals. I am simply sharing our experiences with them. I like to write in a down-to-earth fashion without the technical words that confuse the reader. I have made the Table of Contents basic and you may use this to reference a specific chapter or story. The book also has a full glossary, index and many resource pages you can benefit from.

My husband Ron and daughters, Shana, Amorette and Kyra Jean, and I live in the mountains on Just-A-Little Ranch. It's a beautiful spot 27 miles east of Tonasket, Washington. If we drive to town to pick up a carton of milk the drive takes forty-five minutes *one way*. Our home is private and peaceful. We love it here.

Our family has four breeding sows and one boar to service them. We breed all four sows for spring piglets and half the sows for winter piglets. In the area we live there isn't a problem selling all of the babies. The pigs keep us pretty busy.

Kyra Jean collects the chicken eggs every day. Amorette is raising her own ducklings. Shana is raising her own dog. All three of our girl's bottle feed the calves we buy to raise for meat and breeding. And Ron and I are raising milk goats. Our goats need to be milked once a day. We use the milk for our family use and to offset the cost of pig grain.

The biggest issue I would like to spread from nation to nation if I could? Take care of the animals! Good care. Don't just toss them some feed and water and walk away. Talk to them. Be nice and caring to their needs. You will notice that if you take good care of the animals, they in return will take good care of you. I like to tell anyone who will listen this: Happy animals produce quality livestock.

With this many animals there is always a story to tell! I write a quarterly newsletter for my pig customers and anyone who likes to read about animals. This keeps them updated on the progress at our homestead, as well as new things and stories I learn from other farmers. I read everything I can get my hands on about farming! If you would like to receive a FREE copy of my newsletter, just write your request or email to the address below. Or subscribe from the order page at the back of this book.

It is our goal to become more self-sufficient on our homestead as time goes by. I enjoy hearing stories from readers about their experiences with their animals! Please write me anytime. All letters are acknowledged.

Jeanie Whiting
P.O. Box 1516
Tonasket, WA 98855
Email: Jeanie@foxmtnpublishing.com

Contents

- Introduce the happy couple
- Was she *really* bred?
- Use caution for 14 days after breeding
- Gestation
- 1st Three months
- Last month of gestation
- Second belly stage
- 2-wks before due date
- 1-week before due date
- Sow while farrowing
- Human intervention

ILLUSTRATIONS

CHARTS

Farrowing Charts

Health Charts

Disclaimer

This book is designed to provide information about the subject matter covered. It is sold with the understanding that the publisher and authors are not engaged in rendering legal, accounting or other professional services. If legal or other expert assistance if required, the services of a competent professional should be sought.

It is not the purpose of this manual to reprint all the information that is available, but to complement, amplify and supplement other texts. For more information, see the many references in the Appendix.

Raising animals is not a get-rich scheme. Anyone who decides to raise animals must expect to invest a lot of time and effort without any guarantee of success.

Every effort has been made to make this book as complete and accurate as possible. However, there may be mistakes both typographical and in content. Therefore, this text should be used only as a general guide and not as the ultimate source of raising animals.

The purpose of this book is to educate and entertain. The author and Fox Mtn Publishing shall have neither liability nor responsibility to any person or entity with respect to any loss or damage caused or alleged to be caused directly or indirectly by the information contained in this book.

If you do not wish to be bound by the above, you may return this book to the publisher for a full refund.

Farm Animals
Your Guide to Raising Livestock
is dedicated to my Grandfather
David Byron Burris
1918-2001

1
Brief History

If anyone tells you that a life with pigs, goats or cows is always the same and there are never *any* problems, then they are filling you full of crap!

Every single day is different on our farm. If you have a problem, even the most simplest of all, then your day has not been easy. 'Easy' would be feeding and watering the animals and then going back inside the house to sit down and drink tea.

A Little About Us!
Ron and I decided to move away from the West Coast (Seattle, WA area) seven years ago. Both of us wanted to raise our two daughters in a down-to-earth honest place with lots of benefits for their futures.

Ron has his own business and contracts out of the house and I enjoy the agriculture business; it keeps us both busy. I spent some of my early years on farms and wanted the same for our kids; fresh air and wonderful experiences with nature.

For Ron and me, it started with emu birds, baby chickens and the joy of watching these animals grow. We decided to find a profitable farming adventure and start a business we could eventually retire on. Then we started raising a pig every year to butcher for the freezer.

I fell in love with the hogs we raised. After three years of buying a piglet to grow-out and butcher we bought a female to keep for a breeder. Rosie. And here we are, seven years later.

Ron, Jeanie and Kyra-Jean

Now we have four sows bred twice a year, one breeding heifer (cow), three Holstein steers for butcher, three milking goats, twenty or so chickens, two ducks Amorette hopes will lay eggs and two dogs. Recently we actually acquired a cat!

We have a waiting list of buyers for our baby animals nearly a year prior to the adults being bred. It is wonderful to live in the country and I hope to only *visit* the city for the rest of my life!

Who knows what farming will bring? Ron is still working full-time away from home to keep the bills and mortgage paid. I work full-time writing, home schooling our daughters, taking care of the homestead and keeping life on track. It works for us.

Crazy Ideas We Have Tried

Did I mention that each of our extended families feel Ron and I are crazy? It's because we live a long 45 minute drive to the closest town, our home is heated with wood heat only (we don't even have a furnace if we wanted to use one), and we keep trying these crazy animal ideas! I'll explain.

Upon leaving the West Coast, Ron and I had wanted to start an emu bird ranch. We honestly did. Both of us wanted to become more self-sufficient with this adventure. And although we gave it a fair try, there wasn't a market out there ready for us. We gave the birds away and tried our next adventure.

Next were the horses we felt we couldn't live without. For the horses, we built an expensive three stall barn with a tack room and bought enough tack for two horses to ride side by side; along with many extra things we didn't need. This adventure was short lived and cost us many thousands of dollars. To us, horses are money munchers and we can live without them.

I wanted to raise a few odd animals (the tortoise and camel) just for fun to see if we could do a good job. We didn't get to try running a tortoise rescue farm or raising a camel because Ron was a bit more adamant about not wasting our money after the horse adventure.

Next we tried rabbit farming. We bought two does (females) and a buck (male). And it was awesome seeing those little babies arrive. At first the rabbits were supposed to fill our freezer with meat, but our stomachs couldn't let those cute creatures die! We found out quickly that there isn't quite the market we'd have liked to see for these critters though.

Then of course we bought a cow. You'll read all about her later in this book. But we didn't buy a lot of cows like I wanted to because Ron likes to start slow...and I've come to the conclusion that this is the right thing to do!

Okay. I thought we should buy a zillion day-old chickens and turkeys once, so we could have a huge bird farm and sell our poultry to folks! That didn't even get started...then I wanted pigs.

As you read more you'll see how the pigs have worked for us. I feel they offer a good piece of income to our homestead, as well as a source of meat, and we'll always raise a few pigs because I enjoy having pigs around.

Finally, Ron and I both wanted to start a goat venture. We bought three milk goats and started getting enough milk to supply our family and supplement the piglets which needed extra help growing.

I made cheese and bread with the whey from the cheese. We were making some income here, too. When demands got too great for my book writing career, we sold the goats and decided when things slowed down we'd start up again. And we will next spring!

I am still wanting to start raising bee hives and sheep...but don't tell Ron!

Shana: A Veterinarian

Of course our children change their minds yearly as to what they wish to be when they grow up. Shana is no different. When she was six-years-old she wanted to be a teacher. This wish stayed with her for years. Shana is wonderful with younger children and kids of all ages flock to her.

Around the year Shana turned eleven she decided to be a doctor or a lawyer. Shana wasn't deterred by the commitment she would have to make. She still desperately told us day after day that she would be a doctor or a lawyer when she grew up.

This past year, as Shana is sixteen now, she has once again changed her mind. At the first of the year she wanted to be a psychologist. Shana felt that with all of the counseling she gives her teenage friends on the phone (hours each night) that she will more than qualify for the job. We encouraged her to proceed.

But later in the year she told us she wished to become a veterinarian. Shana has even had a poem she wrote published in a book of art. (Maybe she'll be an writer?)

Shana Whiting, 16-years-old

I hope that as our kids get older they get closer to the actual employment field they wish to be in as an adult. We all feel Shana would make an excellent veterinarian in the farming community we live in. There certainly are many animals on our farm for her to learn a few different procedures.

Shana has helped me give three piglets a daily shot of penicillin. She catches a piglet and holds it while I give the injection. It is great experience for her. And me!

You will learn how Shana castrated her first piglet later in this book.

Amorette: A Doctor

Who knows what a child is thinking half the time? Amorette is like that. She keeps her thoughts to herself. However, if she is thinking something terrible you can tell by the look on her face! This is typical for a fourteen-year-old.

Amorette Peck, 14 ½ years old

Amorette used to collect many rocks. I would find bags, and I mean *grocery size* bags, full of rocks in her closet, under her bed and on shelves in her room. She liked every color, size and texture of every rock. When I went on vacation once I had a picture of me taken next to a huge boulder just for Amorette. So of course it made sense to hear her say she wanted to be a paleontologist for about five years of her life.

When Amorette was around the age of eleven she decided she wanted to be a doctor and this has lasted for the past three years. Amorette did mention something about being a marine biologist recently! Who knows?

Amorette spends a lot of time traveling to her dad's on the coast and he is teaching her to drive a car. Jeff Estabrook is a wonderful dad who spends time with his daughter every chance he gets.

Couch Potatoes

This was a hard time for us. The kids and I are not accustomed to the TV life any longer. It had been three years since last having TV services. Every time I visited with Ron's parents I would complain about not having television.

Well, Ron felt guilty for having TV channels in his RV during the week while he's away from home working. He called the satellite boys and they came to hook us up. After the 'boys' left there was an ugly satellite dish on the side of our home and wires stapled to the house along the deck, but it didn't matter because I wanted to watch 'Friends' and 'Days of Our Lives'. However, it was a luxury which was short lived.

Owning the satellite service for two months opened my eyes to many things. For one, it is absolutely true that people get sucked into the tube and things around them suffer.

Like for instance, chores and the cooking of meals. I sure as heck didn't want to stand in the kitchen and cook dinner when I couldn't get a visual of the TV from there! And who could ask the kids to do it? They were sucked in just as badly as I was. Who's going outside to do chores when their favorite show is on? Not me and not the kids if they could get away with it. With forty channels there was always a favorite show on.

I didn't notice those things right away, though. At first watching the TV was too much fun to notice any pitfalls. It seemed as though we were spending more 'family' time together. Not so. This is just an illusion. When the entire family became sick with the sniffles I opened my eyes to our dilemma.

24

Our family doesn't usually catch colds. We eat pretty healthy and get lots of sun and fresh air. I feel this helps keep sickness at bay and whether it's true or not it works for us. At any rate, we all had the runny noses and sneezing attacks of a winter cold. Not fun. Not fun at all.

That's when I realized we weren't eating as healthy, the chores were suffering and our house was cluttered. I don't like clutter in my home. I feel that as long as your floors are clean, beds are made and the bathroom is spotless, your home is basically clean. But this wasn't the case for us during the time we owned satellite TV. You may be different but we became couch potatoes and ice cream became our best friend!

In the middle of my sneezing attack one morning I decided we needed to start eating better; and taking a walk in the afternoons wouldn't hurt. The girls were disappointed when I called the satellite company and disconnected our services, but things got better the very first day!

The satellite subscription was canceled first thing in the morning. I had to drive to town to buy pig feed and Shana came with me. Usually when we returned home I would find Amorette and Kyra watching a show on the tube but today the house was clean and the kids were outside.

Amorette was cleaning the work room in the chicken coop. We had tossed used feed bags there all winter and she took them out to be burned, swept the room and put tools back in order on the shelves! Kyra was running around outside playing in the fresh air. I asked Amorette what she was doing.

"I was bored without the TV." she said.

Then Amorette proceeded to finish cleaning the chicken coop until it sparkled to her expectations. After this she wandered over to our rabbitry *and cleaned that, too!*

I would prefer my children learn things like animal farming, sewing and quilting, along with cooking and cleaning rather than how to watch TV.

Surfing the Web

Now we are trying the internet. Ron will need to explore employment options for another contract when his present contract expires. Also, I need to keep in touch with updated home schooling possibilities (which are endless) on the internet. The kids (with supervision) can learn things for school such as data needed in reports, facts to help them with their animals and general information.

I have a web site at: www.foxmtnpublishing.com and I sell my books and products in the Store at the website. It is working nicely so far. Check it out. Buy something, I could use the money.

2
Happy Animals Produce Quality Stock

Managing Your Happy Herd

All right, enough about us, let's get down to those animals! Good farm and animal management starts with the breeding stock.

You won't keep a respectable reputation if you sell unhealthy or scrawny stock! Of course, we are only human so there are mistakes made and accidents do happen. We aren't perfect and neither are our children. Ron and I oversee the chores we enlist the kids to do and go to bed each night knowing we are doing our best.

In this chapter I only intend to introduce you to a few things you can do to shorten your chore time and improve your overall herd of animals. There are many ideas for you to consider.

I've heard of a farm with sick and frail animals living in mud and muck up to their ankles. At this farm when perspective buyers came to view the stock of feeder piglets there were dead pigs lying in the barnyard as the folks entered! The barnyard had mud every where and no feed trough or water pans were visible to the buyers.

All of the pigs squealed, trying to get a front row place at the gate, hoping the newcomers would toss them a bite to eat. There was no bedding in the pens for comfort to the pigs, only a thick layer of wet mud. Feeder piglets were not under heat lamps and it was the middle of winter.

Upon being asked, the farmer stated, "My pigs are tough. If they can get through winter with no heat lamps, then they can make it through anything!" He sounded proud, as dead pigs laid behind him in the barnyard.

This was indeed a sad sight to encounter. It shames me that there are still people out there who don't care about the animals they raise; only the money those animals bring them. You can tell a lot by visiting a farm and talking with the owners.

For us, it is important to keep our sows (and all breeding stock) happy which in return will allot for healthy and happy fast-growing livestock.

Manure Management

Are you aware of how much manure you will acquire having animals on your farm? Do you know what you will do with the waste? Here is an average idea of what each animal can produce per year in manure.

Chicken   95 lb. a year
100# Goat   4 lb. a day or 1460 lb.'s a year
500# Pig   32.5 lb. a day or 11,860 lb.'s a year

The Animal's Productive Life

Do you know how old each animal in this book has to be before it becomes sexually mature? Or how long the productive life is for that animal? Here is a basic chart to give you an idea.

First, let me explain something. The average productive life of an animal is not the same as how many years the animal will live. It is an estimated time of the animal's productive life. For example: A chicken will lay eggs for 2+ years, but can live for 10-15 years. Therefore, the productive life of a chicken is 2+ years. Now, here's the scoop!

Animal	Sexual-Maturity	Ave. Prod. Life
Chicken	5-6 months	2+ years
Duck	5-7 months	3+ years
Goat	7 months	10-12 years
Pig	3-5 months	8-9 years

Transporting an Animal

Give some thought as to how you will transport an animal to the vet, the fair or a new home. An old two-horse trailer would be easy to manage for most animals.

Water

Winter can be hard on animals. If you prepare their house for the cold weather and do your best to make sure the animals are comfortable, then your animals will be happier and thus, healthier.

The best advice I can give is to try to have water assessable at all animal locations. The barn, sheds, special housing, chicken coop, etc, should all have a frost-free yard hydrant if possible. Water is one of the main ingredients to life, without water we can't survive and neither can the animals.

Then, place a stock tank heater in the water troughs to keep ice out of them. Having to break ice in the coldest days of winter can be tiresome. For barn animals, you can buy heated water buckets.

Yes, it costs more money to have a stock tank heater but isn't it worth the effort to make the chores easier?

Also, it can save on time and money if you allot one four foot water trough per two housing pens. This will cost you one water trough, one stock tank heater and you only fill one tank a day, *not two!* Just situate the tank so the animals between it can share.

I had some buyers who bought a bunch of piglets one year. They had five or eight children and I thought the pigs would have a lot of company and be spoiled with love at their new home.

We try and spend time with the piglets while they are little and this way they are pretty friendly by five weeks when we sell them. Anyway, the kids had the chore of watering and feeding the pigs. Somehow they didn't do this on a regular basis. It was hot outside and it wasn't long before a piglet died of thirst and hunger.

The children's mom and dad didn't even know why the pig died. I don't think anyone cared because over the next two weeks the other three piglets died from thirst and starvation!

It is extremely important to make sure children are actually accomplishing their chores adequately.

Keeping Grain Bins

Storing grain is important all year long. We use garbage cans with tight lids. This keeps the mice, rats and other predators at bay. And it's a simple method. Use an old 3 lb. coffee can to scoop the grain out with, and keep the chores easy.

What We Do

We breed in late October to get our sows littering in late February, so we do have to take precautions for winter months. Here are some hints and tips:

♥ Keep the pig houses clean and thick with straw so the sows are comfortable and warm.

♥ Set up heat lamps late in fall so if there are low temperatures you can turn them on easily.

We live at 4200 feet elevation and the winters get cold but we haven't had to turn the heat lamps on for our sows until the week they are due to farrow. We turn lights on a week prior in case the sows litter is early.

It doesn't take much more to raise a sow in the winter months if you do it right. Our sows are happy and content and because of this they spend most of the winter lying inside their house, preserving their body heat.

♥ Ron set up a four foot metal trough to water the sows and we keep a stock tank heater inside the trough during the winter. There is no ice to break during winter, which keeps the chores easy.

♥ We have to put a piece of plywood over the top of the water trough so the sows can't get to the wire or break the heater, but that is simple enough.

Flavor in the Meat Cow

When raising a cow for meat, you can raise the cow on pasture. If you add grain to the cows' diet the last six months of pasture feeding, the grain will help add the tasty marbling texture to the meat. Without the marbling in your meat, the meat will seem tasteless, compared to store bought.

Animals Like Company

When buying animals, remember that the pig, goat and cow are social animals. They would rather be kept in pairs.

If you can buy just one pig and one goat or cow, try to pen them so they will be close together. Keeping the animals happy should be the number one priority. Remember, they didn't ask to come live with you, you made the decision for them to come there, so take care to respect their needs.

If raising pigs to butcher size, it is wise to purchase more than one pig. Most neighbors will be happy to buy the extra meat. The pig is the type of animal that wants to eat every time its pen-mate is eating. So, your pigs will grow more quickly and get to that desired butcher weight a lot sooner with a partner.

Straw adds Warmth in Winter

Did you know that six to eight inches of straw bedding can add ten degrees to a cold winter night? It's true! Your animals will appreciate this news more than you do.

Keep Records

It is important in any size venture to keep records. You can't remember everything you do everyday for years upon years. We write it all down and track our efforts with an easy accounting program called Quick Books Pro. See chapter 17 on record keeping for more details and examples of records and charts to keep on your farm.

For a report with blank, simple charts, records and sheets you can use right now as you start up your farming adventures, or as you start keeping records, see our resources for Document #275 *Small Farm Records.*

Newsletter

To keep in touch with customers, it is a good idea to send out a quarterly newsletter if time and money allow it. It's a bit expensive but can work well as a marketing tool when you have animals to sell.

Seek Good Advice

When you're on the move to buy a new animal and you aren't familiar with checking the quality of health or conformation of the type of animal you wish to acquire, it might be a good idea to hire a veterinarian to help you pick out the animal.

Or, if you have a friend or neighbor who knows about the breed of animal you want to buy, maybe they would come along to help.

Ask the selling farmer if they guarantee their animals health for at least three days. Most do.

Take Your Dog Fishing!

Here's a cute story to remind you to involve your animals in your daily routines whenever possible. If you start when you first bring your dog home, it can be rewarding when the dog starts to learn. There is the occasion during the learning process when we all have to just sit back and laugh...

Every year as Round Lake opened up for the first day of fishing season, Nick Baker and his kids would haul their small boat out to fish. There were always a lot of people fishing on the season's opening day.

Nick had just added a dog to their family the week before and they brought him along. The dog was a yellow lab pup (named Bud), and wasn't obedient trained yet. Nick tied him to the back of their truck at five a.m. and Nick and the kids went out in the boat to start fishing. It was foggy out and you couldn't see very far ahead of you.

The kids and Nick were discussing how many fish each of them would catch when they heard a bunch of ruckus. There were folks cussing up a storm! People were yelling that their lines were tangled on something. Other folks were laughing.

Pretty soon Nick heard the culprit. Bud (his new dog) had heard them out on the lake and decided he wanted to come join the fun! After he escaped his chain, Bud started to swim out to his new family (labs are very loyal). The dog didn't realize what he did as he swam, taking fish lines with him, crossing other folks' fishing lines, pulling one line into a tangled mess with another.

Nick decided with all the cussing he better get the heck out of there so his dog wouldn't find him. He didn't want the other fishers to see that it was HIS dog. All those cussing folks would have someone to blame!

So, Nick and his kids started paddling faster, trying to get farther out into the lake. But Bud found them. With three kids and Nick in the boat, they couldn't pull Bud in with them; so they had to turn around and row their way back to shore, paddling through all those cussing fishers and tangled messes with their heads down in embarrassment, and Bud swimming about ten feet behind them all the way!

Dogs in the Winter

Remember your dogs! Dogs can get really cold, too. These days you can help a lot just by adding straw to their housing and making sure the house isn't facing the North winds.

Remember, if you let an outside dog in the house a lot during the winter because you think it is too cold outside for him, it won't help him. The dog will not grow a thick winter coat if his body is too warm and cozy most of the time.

De-stinking Your Skunked Animal

1 quart 3% hydrogen peroxide
¼ cup baking soda
1 teaspoon liquid soap

Mix together, rub deeply into the animal and rinse thoroughly. You will want to do this procedure outside if possible.

Review

♥ Yes, we use a stock tank heater for the water trough in winter months. We don't have to break ice while doing our chores in the cold weather.

♥ We set up and use heat lamps for our sows with litters. Our sows are comfortable and happy and the piglets get the warmth they require.

♥ And we keep the heat lamps on until the temperature is at least fifty degrees at night. Piglets will die easily if temperatures drop below fifty.

If the sows aren't bred too early in fall then it is easy to keep the farrowing in February and March. And finally, we keep the pig houses thick with clean straw for extra warmth and comfort.

Give Your Animals Respect and Love

We try our best to raise healthy, happy, fast-growing animals. One thing I know for sure? Animals can tell if you love them. And if you take more than a few seconds a day to fill water and dump grain into the pails, then you will see what I mean.

If your animals are happy and healthy, they will produce nicely for you. If you do your best and still get poor product, try buying another animal of the same breed and see if you get different results.

♥ No matter how well you raise an animal, genes are hereditary and you will have to cull the bad stock. That just makes good business sense.

At our homestead we treat each animal with the respect it deserves until the day it is sold or brought to the table. I am proud of our animals and I guarantee their health when they are sold. I have a list of folks to call in spring and sell baby animals to even before the breeding stock have been bred.

At any rate, keep your breeding stock happy and you'll get healthy, happy babies. It is so much fun to watch all the babies playing in the yard.

If you use this simple philosophy of giving your animals respect and love, you will be rewarded with happy, healthy animals. It's well worth your time!

What are the basic ingredients to use for the love and respect of your animals?

- ♥ Make sure each animal has clean, fresh water daily. You wouldn't drink dirty or defecated water, why would your animals; unless they have no choice?

- ♥ Feed each animal twice a day, especially pigs. Feeding twice a day keeps the animals from over-hunger which can cause them to tramp most of their feed into the ground as they eat; wasting the bulk of it, and still leaving the animal hungry.

- ♥ Don't forget to reach out and pet your animal! Geez, don't you like to be touched? So does your animal!

Let the animals have a nice, loving and gentle upraising and you'll feel better for doing your best in a world full of neglect.

3
Chick Basics

You don't have to own a lot of animals to have fun. If I were going to recommend anything it would be for you to start slow. Buy one or two animals, see how that goes. If it's enjoyable and it brings you pleasure or income, then build on it. But if it's taking you under financially, don't try and stick it out, back off and try something else.

Ron and I use the above mentioned method on our homestead. Financially, it sure does keep you out of trouble. We have to be careful about the adventures we step into!

A Word about Chicks
Chick – Newly hatched

Day-old Chicks

Our first year here we purchased twenty-five one-day-old chicks from a hatchery through the mail. It worked out fine. The chicks arrived at the post office in town and when they called to say the chicks had arrived, I drove down to pick up our new babies. Every chick was healthy and survived the trip.

Setting Up

When setting up for one-day-old chicks you have to consider all your options. You could use the expensive method and buy everything you need from a hatchery magazine, or go to the feed store and pay a little less for the essentials you need. Or you could always get creative and use what you have laying around the farm!

Housing

Day-old chicks require little for housing except it is important to keep any drafts out of the enclosure. See *housing* in adult chicken section (later in this chapter) for more details concerning housing for older chickens.

- ♥ Day-old chicks need half a square foot per chick until one month of age. After that, one square foot per bird is sufficient.

- ♥ Most important for your chick house is good ventilation without drafts and safety from predators.

- ♥ Put together a 2' x 2' box and attach a heat lamp to be lowered down into the enclosure for the first few weeks.

♥ Set a thermometer inside the box to regulate the temperature.

Then move the chicks to your chicken coop when they are big enough. Keep them separated from any older chickens for a couple of weeks to prevent fighting. A completely separated 3' x 10' room inside your chicken coop, protected from older hens will suffice for a few weeks.

Heat Lamps

♥ Chicks demand to have a temperature of 90-95 degrees for the first week of their lives.

♥ Drop the temperature by five degrees each week until 70 degrees and then the chicks shouldn't need anymore heat.

We got out one of our heat lamps and clamped it to one side of a 2' x 2' box. By setting a thermometer inside the box we made sure the heat lamp was at the correct temperature.

Litter

One inch of wood shavings, rice hulls or ground cobs will make a nice litter. Do not use cedar chips, sawdust or treated wood chips.

Warning

Too much light, heat or over crowding can result in cannibalism because of the natural tendency of the chicks.

Water

A small poultry waterer, purchased from the feed store works great for baby chicks. Our waterer holds half a gallon. You can also make your own with an upside down coffee can with a small hole cut on each side and set upside down inside a cake pan.

♥ Give chicks fresh water any time they need it and check it regularly to be sure they haven't soiled it.

♥ I add electrolytes (a vitamin purchased from the feed store) to the chick's water once a day for the first two weeks. It helps them get through their beginning days of life easier.

Half a gallon seems to be the right size waterer for twenty five chicks because the chicks are so messy that you have to change the water often.

Feed

One-day-old chicks eat more than you think! It is important to check on their feed a few times a day to make sure it is sufficient. The babies run and play and in the process they will step in and drag their food around the box. Chicks create a lot of waste.

♥ The pan to add feed to can be any size, but you don't want the chicks to stand in it. At our homestead we fill a rabbit dish (3" x 5" and 2" deep) twice a day.

♥ We feed our chicks a poultry starter from Tonasket Feed Store. Chicks eat a lot and require feed in front of them at all times.

42

If you plan to do a great deal with poultry then I recommend reading *Storey's Guide to Raising Poultry* and see our resources for hatcheries.

Warning for Parents

♥ Without knowing any better, young children can hold a chick too tight and squeeze it to death.

Be sure to keep an eye on your little children. It is important to let the kids get involved, but they need to be supervised with baby chicks.

Roosts for Chicks

Day-old chicks, of course, don't need a roost. After they have reached four-weeks of age you will want to install a roost inside the brooder (chicken coop).

♥ At four-weeks of age make the roost one inch off the floor in the brooder. Raise the roost height by one inch a week, until six-weeks old.

♥ At six-weeks of age you will want to move the chicks to the chicken coop area. See adult chicken section for more details.

What We Do

Ron and I use the above mentioned box for day-old chick's the first two weeks.

♥ At two weeks of age we move the chicks to a separate area in the chicken coop so they are protected from the older hens.

♥ By four weeks of age we can let the chicks into a rabbit cage and put them in with the hens to get acquainted through the cage.

♥ Then at six weeks of age we let the chicks in with the hens. There is some pecking order but things quiet down quickly.

Headless Chicks

You would think month-old-chickens could be safely brought into the chicken yard, right? Not always so. We don't have a cover over the top of our chicken yard and flying predators can easily enter.

What we do is keep the chicks inside a rabbit cage so they are safe from the older chickens that will 'peck' on them (pun intended). After a time of three weeks to one month goes by and both the chicks and the older hens have gotten used to each other, we let the chicks out of the rabbit cage. There will be some pecking and chasing each other around but usually all is well relatively soon.

But one day something terrible happened. My (then) three-year-old went out to collect the eggs and came back crying. Kyra had found her turkey, Tree Branch, lying in the chicken yard with his head removed and blood spilling out on the ground. It was pretty gruesome. You should have been there to see Kyra's tears and those huge eyes filled with dread.

Here's what happened. The kids and I had let the chicks out of the rabbit cage so they could peck in the yard with the hens. During our absence something had gotten into the coop yard and killed the baby

44

chicks. The older hens were huddling in a corner inside the coop house.

Kyra had raised her turkey, Tree Branch, from a one-day-old and now he lay dead in the chicken yard. That wasn't all. Whatever killed him had left him headless! There were nine other one-month-old chicks missing. Four, including Tree Branch, were lying on the ground, but whatever did this took the nine others away, body and all.

At first I thought it was a weasel that killed the chicks but then I realized that a weasel will suck the blood from their victim, they don't tear the head off. Shelly from Tonasket Feed thought it might be a family of raccoons and said she'd lend us a live-trap to catch them. But we didn't even get the time to pick up the trap.

The next day I was watching the chicken coop from the living room window and I saw a baby chicks wings flapping around. There was a huge black raven hopping around next to it.

"Something's happening at the chicken coop!" I yelled.

Shana, Amorette and Kyra came running with me to check it out. Now we had another eleven chicks missing, and the one I saw flapping around on the ground was still there breathing his last breaths. We were pretty much out of chicks now. There were the six older laying hens but they were hiding inside the chicken coop house. Dang chickens.

Being upset as we were, we cleaned up the mess and tried to forget the whole incident. I didn't know what

45

was killing these chickens and eating their heads off. Ron had to work and when he came home a few days later we took him down and showed him our pitiful chickens. What a waste it was to feed those first 20 chickens and Kyra's turkey for an entire month and then loose them to some varmint.

We realized it was the ravens a few days later. There had been a lot of big ravens flying down into the pig pens (which is right next to the chicken coop) stealing their grain when we fed. I remembered seeing the raven hopping around by the dead chick the other day.

Ron got his shotgun out and shot a good sized raven. Then he draped it over a compost pile near the pens and hoped it would scare the others away. It didn't work, but we didn't have anymore chicks for the ravens to eat so they didn't bother the chickens again.

This was the only incident like this we have had in seven years of raising chicks.

Adult Chickens

I am not an authority on chicken care. We've raised a few chickens for years and this is how we do it. For more extensive advice on chicken care, please see our resources for the book titled, *Storey's Guide to Raising Poultry.*

A Few Words about Chickens
Capon – Castrated Male
Clutch – Group of eggs
Cockerel – Immature male

46

Hen - Mature female
Pullet - Immature female
Rooster - Mature male

Housing

The above mentioned book gives wonderful building ideas, but here are the basic's to get you started.

Necessary ingredients for chicken housing are: Correct amount of floor space, good ventilation without drafts, adequate lighting and safety from predators.

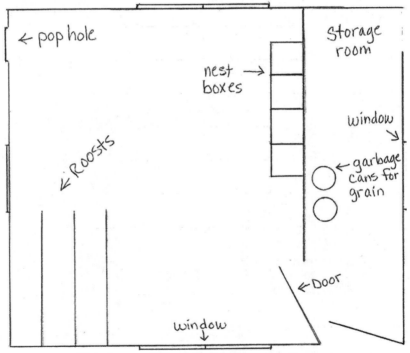

This is our 10' x 12' chicken coop lay-out

When you move six-week-old chicks to the coop they require three square feet of floor space per bird.

♥ Depending what building you have to work with, the coop should have one square foot of window space per every ten feet of floor space.

Optional: Lights would make life more pleasant. Frost free hydrants are good to have close by for filling the waterer's.

Nest Boxes

Chickens start laying eggs at 5-6 months of age and you will want to have nest boxes ready.

Each chicken will lay an egg at a different time of the day; therefore, a large amount of chickens will share a few nest boxes.

Our slant--roofed nest boxes

♥ For every four-five chickens you will need one twelve inch square nest box.

♥ Slant the nest box roof to keep chickens from roosting on it. See picture above.

Roost

At six-weeks old your roost for chickens should be three inches off the ground.

When chickens are approximately three months old you can raise the roosts to normal height.

♥ Normal height for full grown chickens is 24 inches off the ground. You can make your roost set-up like a tier and have the first layer start at 12 inches, etc, until the highest roost is 24 inches.

♥ Four rows of 4' long roosts will be adequate for up to 16-20 chickens.

Fencing

It depends on your financial situation how you will fence your chickens in an outside chicken run. It is not a requirement for the chickens, but it will keep the inside of the coop a lot cleaner if the chickens are allowed an outside yard.

♥ The least expensive fencing is called 'chicken wire' and as long as you dig it in a foot deep around the outside parameter to keep predators from digging under the fence, it is adequate.

♥ You can also use this chicken wire to create a roof-top over the outside pen to keep hawks, ravens and other predators at bay.

Free Range

Free-range feeding is the best you can do for your chickens, as long as they will not be attacked by predators.

Just let them outside of their chicken house during the day; or even the last few hours before evening. The chickens will get goodies from your lawn, as well as save you one third or better of your feed bill.

Remember, chickens can be pests if they escape and get into your neighbors garden, flower bed, etc. Make sure the chickens are safe in your own yard.

Contrary to some beliefs, it is not possible to see a grasshopper when cracking open a chicken egg!

Keeping a Rooster

Hens will lay eggs you can eat whether there is a rooster present or not. But you will need the rooster to fertilize the eggs if you wish to hatch your own chicks.

Rule of Thumb

One rooster will service up to twenty chickens.

Water

Much like all livestock, it is important to keep fresh, clean water in front of chickens at all times. Lack of water can cause disease and death in chickens. The hens need lots of water to develop healthy, eatable eggs.

♥ Feed stores sell waterer's which hold one-five gallons of water at a time. We recommend these watering containers to make sure the chickens do not run out of water during the day.

♥ 100 chickens can drink up to four gallons of water a day. Make sure they have it available.

Feed

Commercial feed from your local feed store is the easiest way to complete the diet chickens require.

We buy 50 pound bags of 'pellet layer' for our laying hens from the local feed store. This commercial blend made especially for egg laying hens has the scratch, oyster shells and grit that the chickens need to make solid eggs.

The chickens don't waste as much feed with the pellet blend because pellets are easier for them to see than a crumble mesh.

♥ Feeders are sold at feed stores and hold from five pounds up to thirty five pounds of feed at a time.

♥ Don't over-fill the feeders or you will get a lot of waste.

♥ Chickens will jump up on the feeder, and tilt it sideways which could dump out considerable amounts of feed. This is a huge waste.

♥ Make sure your feeders are placed at beak level so there will not be such waste. Hang the feeder from a rafter using a rope to keep it suspended in the proper position. Check as the chickens grow to be sure it is still at the right height.

Storing Grain

It is a good idea to keep your grain of any kind in rat-proof containers. This way the mice, rats and other critters cannot get inside and open bags, etc.

♥ Set up extra garbage cans with tight lids to store the feed inside.

Laying Eggs

The biggest reason to keep chickens, besides eating them, is to collect the eggs to eat.

Fresh eggs have a completely different taste than store bought eggs. You will notice a difference in color and taste the first experience you have with fresh eggs, and possibly never buy store bought eggs again!

Hens start laying eggs at five-six months old. First year chickens lay about one egg a day, and this decreases as they get older, but the eggs get larger as time goes by.

♥ Chickens can lay between 200-280 eggs per year for the first two years, depending on breed. You can expect two eggs a day per three hens. The number of eggs laid drop dramatically in the third year.

♥ Egg laying chickens require 14 hours of daylight during the short winter days. A hanging light set on a timer in the center of the coop will provide nicely for this necessity.

Checking for Fresh Eggs

When checking for fresh eggs, put all the suspect eggs into a bowl of water. Fresh eggs contain little air, so they will sink.

Molting

At 18 months of age, usually during winter months, chickens will drop their old feathers and re-grow new ones. This is called molting.

♥ During their molting, chickens usually do not lay eggs. Some do, but the majority won't.

♥ Chickens molt from between 12-24 weeks.

At our homestead, we get two-three eggs a day during winter and this is with a light bulb in the coop so the chickens receive 14 hours of light per day.

A bunch of ducks are called a bevy

4
Duck Basics

Ducks are hardier than turkeys or chickens. Ducks require less attention and they grow much more rapidly. Also, ducks will augment most of their diet from free ranging, if let out to roam.

Ducks are easy to care for. They need little space and nearly no housing. They do not require a pond or a brook, although they will be pleased if you provide either.

Did you know that ducks have personalities? Frolicking ducks on a pond or in a pool of water is a pleasure to see.

My daughter Amorette has enjoyed watching her two ducks splash and dive and take baths in a plastic infant pool she provided for their pleasure.

A Word about Ducks
Duck - Mature female
Duckling - Newly hatched
Drake - Mature male
Bunch of ducks - Bevy

Amorette's Ducks are the only ducklings we've raised. She chose to buy the white Pekin breed. We picked out two of them hoping one would be a male and the other a female. This didn't happen. Amorette has two males but she loves them both and won't give either up.

Buying Baby Ducks
Amorette bought her ducks at a week of age from our local feed store; Tonasket Feed Supply. It is also possible to order from a mail-order hatchery but you are forced to buy a large minimum number of birds.

If you order through a hatchery, you could ask close neighbors if they would like to buy ducks, this would make the bulk purchase cheaper and the number of birds you get could be exact.

It is not hard on the ducks to travel to your home from the hatchery. When ducks are hatched, they have 2/3 yolk left covering their bodies (which they will eat) and they do not require any food or drink from you for several days. This is the best time to ship ducklings.

Water for New Ducklings

When you first bring home your ducklings they will need water but may not know how to drink yet. To teach them, place one teaspoon of honey into one gallon of luke-warm water and dip each beak into the mixture. Do not use a dish the ducks can dump over and get soaked with. See 'wet feathers' for more.

Brooding

Young ducks need to be in a brooder. Duckling needs are much the same as for chicks.

♥ Ducklings normally need brooding for four weeks.

♥ A brooder must provide one square foot of floor space per duckling.

♥ They require a temperature of 90 degrees the first week and then drop the heat by five degrees until the temperature outside is 55 degrees or higher.

♥ In the summer weather you can supplement with just a light bulb in their shelter at night.

For 20-40 ducklings you can suspend a 250 watt heat lamp 18-24" above the brooder area. You might want to use two lamps in case the first one burns out.

Wet Feathers

♥ Do not allow a duckling to get soaked by rain before it is 4 weeks of age or until it has feathers.

♥ Ducklings should not be allowed to swim before 6 weeks of age because the oil is not fully distributed on the feathers and the duckling will be susceptible to chills and possible sickness.

Never!

Do not ever try to catch a duck by its legs. You can damage the bird. Catch the duck with your hands resting in front of his neck and gather him into your arms.

Housing

It's nice that housing for ducks is simple. Any enclosure will do. An old shed or a simple hutch with a dry floor and protection from severe weather and predator's will be adequate.

♥ You should allow 5-6 square feet of floor space per full size duck, in the housing area.

♥ South facing doors will allow the sun to enter the doorway during the winter; but make sure this isn't the same direction the wind comes from.

♥ The floor will be less cold and produce a heat through decomposing manure if you keep a deep litter in the housing.

♥ Keep drafts out as best as you can.

♥ Shade must be available to young birds in hot weather.

Amorette and Ron built a shed for her ducks from an old apple bin. It took less than one day to build. They chose to leave a dirt floor inside, which made the project even easier.

58

Ron had some left over roofing and he and Amorette used two by four boards to attach the roofing at an angle to the roof of the apple box, so the winter snow would slide off.

Outside Yard

Each duck should be allowed 10-25 square feet of outside yard space.

Fencing

See adult chicken section for fencing ideas.

Nest Box

Ducks lay eggs and therefore you will want them to have a nest box in order to keep the eggs clean.

♥ Like chickens, ducks will share a nesting box. Allow one 15x15 inch nesting box for every 4-5 ducks.

♥ Set the nest boxes on the floor, without a roof.

Roost

Ducks do not require a roost.

Keeping a Drake

If you plan to breed your ducks then you will want to buy a drake. One drake will service up to six female ducks.

Note: Like chickens, you don't have to own a drake to receive eggs from the duck. A duck will lay an egg whether there is a drake present or not. You need the drake to fertilize the egg if you wish to hatch it.

The drake will be more colorful than the female. This makes it easy to tell who is who in the yard.

Equipment

Equipment for ducks consists of feed and water dishes as the basics. Amorette provides her ducks with a small swimming pool also, and they use it even in cold weather until it freezes.

♥ Ducks don't scratch at their food. You can use chick feeders or old pans to hold their feed. It's kind of funny to me that ducks eat poultry food. We keep their pans filled with feed at all times.

♥ If you can let the ducks out to range in the yard it will cut your feed costs considerably. Keep their water full and accessible at all times.

Feed

This is just a basic chart for your duck's feeding needs. Refer to more extensive readings for more detail. See resources.

♥ **Starter/Grower** – 20-22% protein for the first 0-8 weeks.

♥ **Developer** – from 8 weeks to sexual maturity.

♥ **Layer (maintenance)** – from 2-3 weeks prior to the egg-laying age of the bird.

♥ **Breeder development** – from two-three weeks before breeding age.

♥ **Breeder** – during the breeding season.

If you do not intend to breed your ducks just use the egg laying maintenance grain your local feed store recommends for ducks.

Ducks can eat just about anything that you do. Before throwing out your table scraps, check with your duck to see if they will eat the garbage.

♥ Vegetable scraps
♥ Stale bread
♥ Any scraps from the garden
♥ Table scraps

Water

Water is one of the most important resources that ducks require. Ducks need fresh, clean water in front of them at all times. A swimming pool cleaned out and refilled daily will meet their needs nicely.

♥ Ducks drink 3-4 times more water than chickens. Make sure they have it assessable.

Warning! A chicken waterer for inside watering needs will not work for ducks because they need to immerse their entire beak when drinking.

Predators & Dangers

If you have dogs or other animals that run loose where the ducks will roam, remember to introduce them to each other.

Predators are the same as for chickens, of course. Watch your dog the first week or so to be sure he isn't interested in tasting your duck.

The health of your duck should be just like all other animals. Prevention is easier than treatment.

♥ Keep the duck house and yard clean. Check frequently for garbage the ducks might hurt themselves with, like nails, screws, etc, while investigating their yard.

♥ Keep all ducks away from stagnant water.

♥ Remember to keep young ducks from becoming chilled, or immersed in water before six-weeks of age.

♥ Avoid over-crowding ducks.

5
Rabbit Basics

Rabbit's are another simple creature to care for. Some people like to make rabbits pets and others keep them for a prolific meat source.

Note: Rabbit's are very prolific. One ten pound doe can produce in her litters up to 120 pounds of meat per year!

A Word about Rabbits
Buck – Mature or immature male
Doe – Mature or immature female
Kindle – Give birth to a litter of rabbits

If you are going to raise a few rabbits for butcher or just keep them as pets, remember to take good care of them. Rabbits have feelings too.

Rabbits don't take a lot of care but they do need daily tending. Keep water in their cages at all times. Rabbits can die without sufficient water to drink. Also, they will need a salt lick.

Housing

36" x 36" is a good size hutch for any rabbit. Smaller rabbits just have more room in this standard size hutch. You can find used cages for next to nothing at rural garage sales.

♥ The hutches should be placed away from falling rain and snow, windy conditions and neighborhood dogs.

♥ You can hang a self-feeder and water tube bottle from this wire cage easily.

♥ Even the salt lick which rabbits need will be simple to attach.

♥ Wood chips, sawdust, or peat moss under the hutches will absorb moisture and minimize odors.

♥ Rabbits don't require sunlight so they can be raised inside a barn or garage.

A good book to read if you would like to know in-depth directions on raising and breeding rabbits or building a hutch yourself, would be *Storey's Guide to Raising Rabbit's*. See resources.

Feed

Rabbits are only fed once a day. They require very little in the way of rabbit pellets. Check with your local feed store to see what is recommended in your area.

If you feed the commercial rabbit pellets, you can't go wrong with their dietary needs. Rabbits require alfalfa hay, phosphorus, calcium, trace minerals and salt. Commercial pellets made for rabbits will have all of these ingredients.

Using a self-feeder and water tube bottle is a simple way to meet the rabbit's needs.

♥ Two-nine ounces of rabbit pellets a day are sufficient depending on the size of your rabbit. Big breeds require the larger amount.

Ounces of Pellets to Feed

	Bucks	Does
Small Breeds	2	3
Med. Breeds	3-6	6
Large Breeds	4-9	9

♥ Alfalfa hay is put in cages to keep rabbits occupied because they like to chew. It's a good supplement in cold winter months.

Greens & Veggies

Did you know that it actually isn't good to give lettuce to rabbits in large amounts?

You can give rabbit's carrots, cabbage, lettuce and other garden scraps, but not in abundance. Do some research before feeding your rabbit too many greens.

If you have a nursing doe, she will appreciate a nice slice of juicy apple!

Water

Rabbit's require water in front of them at all times. Without water, rabbits will not survive.

♥ Use a water tube bottle which attaches to the outside of the cage for easy filling.

♥ A nursing doe can drink a gallon of water a day to produce enough milk for her growing litter. Make sure it's assessable at all times.

Winter Months

We add a chunk of wood into each rabbit hutch so the rabbits can sit on the wood.

♥ This will deter their toes and feet from freezing to the bottom of wire hutch floors.

♥ Rabbits also like to chew on the wood to keep their teeth in good shape.

Where we live it can get to be twenty degrees below zero and a rabbit could get frost bitten toes or feet and then need amputation. Of course we've never had a problem with frost bite here, but it's better to plan prevention rather than have to treat the problem.

Breeding

The book *Storey's Guide to Raising Rabbits* has an extensive chapter on breeding rabbits. I'm not an authority in this area. We did raise a few litters, though, and the best advice I can give you if you intend to breed (besides reading all the books you can about it) is that:

♥ 3 seconds = 8-12 babies **every** thirty one days!

♥ Does are sexually mature at six-months-old, depending on her size and breed.

♥ If you try breeding without reading more extensive writings on the subject, remember the doe should be taken to the buck's cage for mating. This prevents fighting.

♥ Does will attack a buck when he is taken to her cage.

Keeping a Buck

When choosing a buck remember he is half your herd. Be picky! Good conformation is important.

♥ Choose a buck with a shiny coat, bright eyes, not too fat, with a full and large scrotum and two fully descended testicles for best performance.

♥ Bucks become sexually mature by eight months of age depending on the breed and size.

♥ One buck can service up to ten does. Do not use him more than every 4-7 days for best performance.

Molting

Molting is not a concern for rabbit owners. Rabbits will lose their fur and then re-grow it once a year.

Predators

There are many predators to rabbits. The dog, cat, weasel, snake and rat are the most common.

6
Goat Basics

The goat is a wonderful animal. You can own and raise a milk goat for a fraction of the cost and space requirements of owning a milk cow.

A good doe will produce two-six quarts of milk a day for up to 200 days per year. And if you stagger the breeding dates of your does, you could have milk all year long!

A Word about Goats

Billy or Buck – Mature male
Doe or Nanny – Mature female
Kid – Newborn
Wether – Castrated male

Shelter

Goats are quite resilient. In the wild they will find shelter in the worst of conditions and never need intervention from humans.

As we have domesticated them the goat will need a cover which holds back the rain, wind and drafts; especially when young goats (kids) are present.

Barn

Our barn was initially built for horses. We sold our horses and then it became the cow barn. Now the cows have been moved to the pasture.

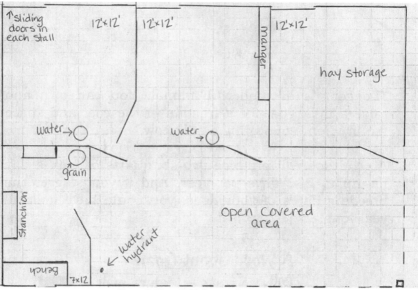

Our 24' x 36' barn lay-out

Our barn has now become the goat barn! The barn has a dirt floor and an eight foot long manger to hold the goats' hay without too much waste. There is a picture below of the manger Ron built our goats.

Each goat should have 12-25 square feet of floor space, depending on the outside space provided.

♥ Two 12' x 12' stalls open into each other to form a local area the goats can hang out in. Goats are social creatures. They want to be together.

♥ The third stall is the same size and is used to store alfalfa hay or to isolate goats which are sick, injured or kidding.

♥ The hay manger is eight feet long. It is simply built with a piece of welded wire stock panel (hog panel) attached to the back wall of the stall. As a floor for the manger (two feet off ground level), we used a ½ inch piece of plywood lumber with a lip at the edge to hold hay in better.

♥ The yard hydrant is just outside the tack room door. It is preferable to have water in the barn for easy care to the goats and also for cleaning and rinsing milk equipment.

♥ We hang the goats' water bucket 36" off the floor, as goats will not drink dirty water.

♥ Our tack room is 7' x 12' and has a concrete floor. We use it to store goat-chow (grain) in garbage cans with tight lids, keep our first-aid kit for the barn animals, and milk our goats. We do have electricity.

♥ The milking stanchion is built so we can remove it when necessary for cleaning.

♥ Finally, our barn has a 12' x 29' covered area which is open on two sides. We use this space to park Ron's tractor. Also, in the winter, Ron can use this covered area to build projects and keep out of the weather!

Manger

Goats will defecate where they stand; whether there is feed beneath them or not. Tossing hay on the ground for goats to eat is silly. The goats will not eat the hay once it is dirty. They are picky eaters.

We keep a simple eight foot manger on a wall within the goat stall. This keeps the hay safe from the weather as well as keeping the waste down to a minimum.

I explained how our manger was built and the materials we used to build it within the 'barn' section. Here is a picture of our manger to give you a visual.

Heat

A heat lamp may be necessary if the winter is harsh. But if it's not extremely cold then you want to let the kids get acclimated to the weather like they would in the wild.

Fences

Fences and gates definitely have to be goat-proof. Goats are ingenious for escaping poorly erected fencing. What kinds of fencing are suitable then?

Well, chain link is best but quite expensive. Stock animal fencing is secure enough but also very expensive; and you would need to install a line of electric wire six inches off the ground.

♥ You need to check to make sure the electric fence is still working periodically or the goats and their babies will escape and could be attacked by predators such as dogs, coyotes, etc.

We like to use a fence charger which has a pulse to it. This way we can hear the ticking sound which says it's working properly. And it is easy to hear if something is hung up on the fence, like debris.

Goats do not need to pasture since they would rather browse your wooded areas and chew up trees and brush.

♥ You can use woven wire (field) fencing at a reasonable cost, but horned goats can get their horns stuck in the squares. This fencing doesn't seem to last long regardless of what you do to make it secure. Run a line of electric wire at nose height to keep the goats from walking on the fence.

♥ If you need a small outside enclosure for goats to receive their food, try using welded wire stock panels (hog panels) attached to metal t-posts with a line of electric wire at nose height along the inside. These panels vary in height; ours are 52" high and 16' in length.

No matter which fencing you choose, it needs to be at least four feet high and will last longer if you run a line of electric wire six inches above the ground and at nose height, along the inside of the fence line.

Goats really just need a small exercise area where you can bring their food to them. An outside yard with 12-25 square feet of floor space per goat is sufficient.

If you add rocks, boulders and climbable objects for the goats, the goats sure will appreciate your efforts. And it will help keep their hooves trimmed. Make sure you don't locate these objects next to a fence or the goats will escape!

Feed

You can pasture goats if you have enough land. We have 40 acres here and it is possible for us to pasture our goats for part of the year.

The rest of the year we have to make sure the goats get enough nutrients and vitamins in their feed to produce the milk we desire.

It takes approximately ¼ ton of grain and ¼ ton of good quality hay per year *per goat* to meet the adequate requirements in feeding.

♥ Allow a milking goat to have one pound of goat-chow (grain) for daily maintenance and another half a pound for each pound of milk produced. Contact your local feed store.

♥ Good quality alfalfa hay should be kept in front of a milk goat free choice.

I provide the grain while the doe is being milked. It gives her something to do while I'm busy milking.

Of course, if you aren't milking your does, then there is no need to give them grain. Just feed them hay as directed above.

♥ Never feed moldy hay.

♥ Goats are very picky eaters. They will not eat off of the ground and won't drink water if it is the least bit dirty.

Your local feed store can provide you with the grain necessary for milking goats.

We mix the goat chow feed with our feed stores cheaper bulk grain. Bulk grain from your area may differ in contents and price from ours, but we didn't find a decrease in milk production with this mix.

For more extensive guidance to the goats' digestive system and nutritional needs, please read *Storey's Guide to Raising Dairy Goats*. It is an excellent book!

Water

Goats need clean, fresh water in front of them at all times. In order for a milk goat to produce milk, she has to drink enough water.

A full grown goat can drink up to 12 gallons of water a day.

When Picking out a Milk Goat...

When picking out a milk goat for production on your farm, there are a few things you should be aware of.

♥ Pick a well formed udder, free of injuries and scabs or extra teats. There should be two teats.

♥ Teats should be a nice even size, not overly large or especially small.

♥ Check to see that the hooves have been taken care of. Some excess growth is okay but make sure the hooves are not shaped like an elf boot.

♥ Make sure the hind legs are not bowed. They need to be a bit wide to support a nice size udder, but not extensive.

♥ Eyes should be bright and clear. Curious.

♥ Hair should be soft and thick, not dull with bald patches.

You will never find a perfectly balanced udder, but avoid a terribly unbalanced set. Also avoid an overly large sized udder or an udder that is pointed sideways.

Psychology

With our four-year-old daughter helping at chore time and wanting to be in the middle of everything, it is important that we protect her. She is innocent as all young children are.

♥ Remember, if you treat the animal well, it will in return treat you well.

♥ When buying an adult animal you should spend time with your goat daily to evaluate the personality and temperament. This is vital on our homestead.

♥ We prefer to buy animals as babies and raise them ourselves, but goats have a docile background and usually want to please you.

♥ By watching the animals walk, play and hang around the goat yard you can see when trouble is on its way.

♥ Goats will show signs of health problems long before they become chronic. Hair will become dull. Goats will lie around the yard more and become less vibrant.

♥ Just keep an eye on the goat yard to foresee problems in your herd early.

Worming

While most goats forage, they consume parasites through the land. Worming twice a year will help clear any internal parasites.

♥ If you choose, you can take a fecal sample to your local vet to examine. Your vet will use a microscope to detect parasites. If let go, parasites can do internal harm.

If no parasite eggs are present in the sample you took to your vet, your vet might have said, "See you in six months."

♥ Since goats gather parasites from the land it is inexpensive if you bring one sample from your herd to your vet to test for parasites. Then you can treat your entire herd if need be.

I plan to worm my milk goats a month before they are due to kid each year. We generally don't have parasites here and this will keep their systems flushed clean.

Grooming

You can use an electric dog trimmer to keep your goat groomed at all times. Basically just keep the hair on the belly trimmed back, and all around the udder should be shaved clean.

♥ Some people say you should brush your milk goat before each milking session. I brush mine every few days and this works fine.

♥ Keep the udder area shaved and clean. This will help with hair and debris falling into your milk bucket.

Trim Hooves

You can use a pair of goat hoof trimmers for this job. You won't believe how nicely they work until you try them. It is a good investment.

The outside layer of a goat's hoof grows like fingernails do, and need to be trimmed periodically. Don't let your goats feet look like pointy toed elf boots, as this could cripple the goat.

Keeping a monthly schedule of clipping your goat's feet is best. The goats will begin to remember the routine and not fight back so much when the chore is being accomplished.

♥ Goat's do not like their feet lifted. They are not comfortable balancing on three legs.

♥ Position yourself rear to rear with the goat, pick up a foot and brace the leg securely between your legs. You have to hold the foot firmly. The goat will kick out and you could cut yourself with the hoof trimmers.

♥ Clean out all the manure and dirt caught in the hoof. Now you can see the over-growth.

♥ Carefully cut off any over-growth of nail. You want the hoof to be flat and smooth when done.

♥ Trim off excess toe at the front of the hoof also. This area does not get warn down naturally and needs to be kept trim.

You will want to watch someone doing this job before you perform the task because hands-on help says it all!

Herd Behavior

Goats have a herd order just as most animals do and they will establish this rather violently, if you keep more than one goat.

♥ Even when the others bow to the herd 'boss' she will push, ram and head-butt them a few extra times to be sure they know who is boss! It isn't a lady-like thing to do.

♥ But once everyone in the new herd knows who the boss is there shouldn't be any problems.

Katie was the herd boss at her old farm. And Keave was the boss at her farm, with Kyann under her. When they all came together here I kept them separated from each other through a barn stall. They could see and smell and talk to each other but couldn't get at each other to argue over who would be the boss yet.

I watched their attitudes and soon realized they wouldn't actually kill each other. After three days of introduction through the stall wall, I opened the door to the outside yard and let Keave and Kyann out to meet Katie in the flesh.

Each had to smell the other. There were small growl-like noises. Kyann didn't have a problem backing off and letting the other two establish who would be the boss. Katie was sure to win, I knew that. But Keave didn't and she wasn't going to back down without a fight!

Around the pen in a circle they went. Katie would lower her head and Keave would meet her with a powerful ram to the head. Their heads must have been sore after twenty minutes of this.

Katie and Keave would then start the process over with a variation. Now Keave would raise her front legs off the ground before meeting Katie (who would in turn rise off the ground) to give her a good ram to the head!

It wasn't long before Keave backed down and Katie became the boss of our homestead. I'm just glad none were hurt in this behavior process.

Castration

This procedure is much simpler than the process with piglets.

I have an elastrator. This is a tool which you put a rubber band on and it stretches the band into a large circle so the testicles can be lowered into the circle.

After the testicles are in place simply close the elastrator, the rubber band slips onto the male and your job is done.

♥ Of course it's not over for the goat and he could probably tell you a few things if you spoke the same language!

♥ The goat will run and lay down a lot. He might cower in a corner for half the day. Or he may be as normal as the day he was born. But I doubt the latter will happen.

Pack or Working Animal

- ♥ If you plan to train kids you need to start as early as possible.

- ♥ Goats need to be accustomed to humans from the very start of life.

- ♥ With bottle feeding they will soon feel you are their parent no matter who you really are.

- ♥ Their thinking of you as a parent is important for trust, respect and devotion later on when you are out on a trail with the animal.

After our milk goats have kids next March I will pick an exceptional female and train her to pack for my husband. Ron wishes to have the goat pack his supplies and any essentials he feels he needs while hiking in the mountains around us.

By picking a female I am hoping to kill two birds with one stone. If our animals aren't pulling their own weight, then they aren't producing for us. All animals have a purpose and even though we love them each dearly, we have to keep the business aspect alive to keep our homestead thriving.

The pack goat will be two things for us. She will be a pack goat, of course, and she will naturally be a milk goat. I'm assuming Ron will rarely hike with a goat and this way she isn't kept around unnecessarily (eating food and costing us money) in between hiking trips! Plus a goat can be trained to pull a harness, a cart, and till a garden with a special cultivator made for working goats, etc.

Basic Milk Equipment

Milk Room

My milk room allows me to accomplish the task of presenting clean, fresh goat milk to my family.

Our milk room is a seven by twelve foot room in our goat barn. Ron built me a portable stanchion milk stand. It works great. If we relocate the milk room to another building, I can take the stanchion with me!

Stanchion

Ron got the directions for this stanchion out of a new favorite book. The rewrite of *Raising Your Milk Goats Successfully* is now called *'Storey's Guide to Raising Dairy Goats'* and is a wonderful source of information. There are plans with a material list to build a nice stanchion which folds up to the wall when not in use. Borrow this book from your library or buy it to build your own reading resource library.

I noticed our goats' hooves collected fecal matter rather quickly and I wanted a second stanchion to use for grooming chores before entering the milk room.

♥ While I'm milking the goat I could think of twenty things I would rather see (and smell) than the crap in between their cloven hooves!

83

♥ With the grooming stanchion I could keep the goat intact outside of the milk room while I brush away loose hairs and clean or trim the hooves if needed.

Necessary Milking Equipment

Of course, we were just starting out and, initially, we wanted to buy *only* the essential products. There are a few start-up costs, but you don't have to spend a lot of money to get milking. Be creative and use what is lying around your kitchen if you need to.

You can keep things pretty simple by using a kitchen bowl to milk into, creating a milk strainer from two inexpensive funnels, and storing the milk in fruit jars. But if you are going to milk goats from 365-730 times a year, you might want more satisfaction and better quality from proper milking equipment.

♥ **Goat Milking Pail**: I opted to pay more and invest in a goat milking pail. The pail is stainless steel and comes with a covered hood which protects the milk from falling hair and debris.

♥ **A Strainer Cup** is necessary for the first few squirts of milk each time you milk a goat. These first squirts of milk are full of bacteria that rest just inside the teats.

♥ **A Larger Strainer** is necessary for clean milk. Use this strainer to drain the milk through after your job of milking is complete or your bucket is full.

- ♥ **Jars to store your milk:** It is important to find a few quart size glass jars. You can store your goat milk in these. ½ gallon size jars seem to be optimal.

- ♥ **Tight Jar Lids:** *Hoeggar Supply Catalog* sells handy plastic lids which fit the quart or ½ gallon jars perfectly. No leaks and they are easy to keep clean!

- ♥ **Baby Wipes:** Baby wipes are perfect to clean the teats, udder and your own hands before and after each milk session. Then use a teat dip diluted spray before sending the goat back to her pen.

- ♥ **Teat Dip:** Use teat dip to clean each teat after milking the goat. This closes the milk tissue until you milk again, so bacteria can't get into the udder. Dilute this dip and put it in a spray bottle for easy use. It is simple to spray the teats with a spray bottle.

- ♥ **Other Cleaning Supplies:** There are special cleaning supplies necessary for the milk equipment to keep away the milk build-up which happens when using stainless steel products.

Should you wish to take a gander at the catalog, call *Hoeggar Supply* at 1-800-221-4628.

We bought powdered dairy soap to wash the equipment in the kitchen sink with, then an acid detergent to use once a week for deep cleaning. This way we keep the milk clean and fresh and won't get any milk build-up which carries bacteria.

Do not feed the buck alfalfa hay!

7
Keeping a Buck

Milk goats must be bred yearly to keep producing milk. If you have one or two milking does then try to rent a neighbor's buck once a year, as this will be a lot cheaper than keeping your own buck. A buck requires the same expenses, such as a house, feed, bedding, etc., as a doe. And they do stink!

When deciding if you need to buy or rent a buck for your does, take your personal preferences and the amount of time you have available into consideration. If you don't have the time to monitor each does heat cycle and track the information so you can take your doe to the buck when she's in heat, then you probably want to keep your own buck so you can just open the gate and let him in.

If you have more than a few does and plan to breed them on your farm, then you will need to keep a buck. Here are a few hints.

♥ One buck can service up to 100 does.

♥ Keep your buck at least 50 feet away from does to prevent his odor from tainting the milk.

♥ Choose a buck with superior offspring records and excellent conformation. You want to improve your herd, not just breed them!

The Bucks Shelter

The buck requires much the same housing as the doe, with the exception that you want the structure to be rock-solid. Using 2x6 and 2x8 pieces of lumber will accomplish this adequately. Remember, this 200 pound beast will be pounding his rock-solid head against the shelter walls for fun, and he won't feel a thing.

♥ An especially sturdy 6' x 8' shed will make any buck happy to sleep in.

♥ A buck's outside yard should be at least 6' x 30'. Bigger is better if you ask the buck!

♥ It would be ideal if you could feed and water the buck without having to enter the shelter or pen.

The Mature Buck

One buck can service up to 100 does. Isn't that amazing? Here is the low-down.

♥ Limit the bucks breeding to just a few services before he's one-year-old.

♥ During his first year limit his services to 10-12 times.

♥ After the buck is mature, he can be used to his full capability.

Feeding a Buck

Feed your buck the same as you would a dry doe, with the exception of alfalfa hay. See warning below.

♥ Feed the buck a good grass-hay mixture.

Warning on Alfalfa Hay

Do not feed the buck alfalfa hay. Alfalfa hay has a lot of calcium that the buck cannot digest and he could end up with kidney stones.

♥ Just feed your buck a grass hay mixture to keep him safe.

Water

A buck can drink up to 12 gallons of water every day. Make sure the water is assessable to him.

Early Training

It is important if you raise a buck (kid), to start training him as soon as possible. This buck will grow to be 200 pounds or more and you need to be able to handle him.

A 200 pound buck can be dangerous when does come into heat. He will want to service them. If you teach him respect and use great caution while handling him, he can be taught to listen to you. But you have to start as early as a week-old to achieve such goals. The buck needs to know which one of you is the boss, unconditionally.

For a great book on training goats to be pack and work animals, see our resources for *The Pack Goat*.

8
Preparing for the Goat

Ron's Idea Formed Into My Plan

Once I get a notion inside my head it rarely goes away by itself. I find myself gnawing at the very idea for hours each day, pondering what the outcome could be if we did this or if we did that...And it's maddening for Ron. I drive him nuts! But he sticks by my side and I love him with all of my heart for that. As you may have guessed, my next crazy idea was to raise milk goats.

The idea started with Ron a year or so ago. He wanted to have a goat to pack around his supplies while he hikes through our mountains. I told him he was crazy. Whack-o! When he said we could milk the goat and use the milk for our own consumption I nearly fell over!

"Drink the milk ourselves?"

"Yes," he said, "It would save us some money with all the milk we drink."

Well it took me quite awhile to form the notion into a working idea inside this brain of mine. But now I can see us milking a gallon a day for our own usage, and another gallon or two for our wonderful pigs. Or I can make cheese! I do want to make cheese!

Not sure about Adding Milk Goats

Ron and I still couldn't decide if we wanted to have milk goats or not. The set-up process was pretty expensive.

- ♥ For up to five milk goats you will need a barn of 18' by 18' with room to store hay and include a clean milking room.

- ♥ Milk goats eat approximately three pounds of grain and three pounds of hay daily. It adds up to ¼ ton of grain and ¼ ton of hay per year/per goat.

Since we already have a three stall barn (24 x 36), and it has a tack room with a concrete floor (7 x 12'), we are still pondering whether or not to use the barn for a goat facility. We could use the tack room for the milking parlor. And Ron could easily take down one stall wall to open up a 24' by 12' stall for the two or three does we would own. There seem to be more pros than cons for having milk goats.

Pros of Owning a Milk Goat

Ron and Kyra tried the goat milk from a neighbor and couldn't tell the difference from store bought cow milk. Secretly we substituted the kids' store bought milk at dinner one night with goat milk. All three girls finished their milk without one question or comment!

Of course after telling the girls that it was goat milk and not our regular store bought milk, they insisted that they knew it all along and could tell that something was 'wrong' with the milk. But we think the empty glasses spoke louder!

- ♥ If we could use our own goat milk to supply our household, this would save us the cost of one gallon of cow milk per day; which is what we currently consume.

- ♥ After calculating start-up costs, we found that one gallon of goat milk is about $1.00 cheaper than a gallon of store bought milk.

- ♥ One milk goat will pay for its way here within six months time! With two or three milk goats we could supply our grow-out pigs with lots of milk instead of all the grain we usually pay for.

- ♥ And our sows could enjoy a bucket full of goat milk here and there.

Cons of Owning a Milk Goat

The only con I could think of is the fact that a milk goat (like any animal) will tie you down in regards to responsibility. You can't just quit milking a goat and come back to start again the next week. And the set up cost is expensive.

Our Barnyard

It was a wonderful spring day for cleaning out the barnyard! Ron and I had it all under control. The kids were cleaning the stalls from a winter with cow droppings. Kyra was trying to help but more or less getting in the way. Ron was using the tractor to push the cow droppings out of the outside paddocks. I was making sure the animals were all fed so they wouldn't get in Ron's way.

We wanted the barn and barnyard nice and clean for the goats! Goats can fit into any picture and it doesn't take much to house them but we wanted it all to be perfect.

Ron and I spoil our animals. We feel we need to give them the best care if we expect the best production in return. It's only fair. And it makes good sense. Happy animals are a pleasure to be around.

Terry Dean's Farm

The kids and I visited many goat farms! We wanted to be sure raising goats would be a good thing for us, and something we could handle. Of all the farms we visited, here is a diary of my favorite.

This was our trip to Molson. Ron was 3 ½ hours away in Moses Lake working so he missed out on this visit. Mom wanted to come along for the ride. It was beautiful and sunny outside. An excellent day for a drive. Amorette, Kyra and I picked up mom/grandma at her place in town and we were on our way.

Terry Dean lives fifteen miles out of Oroville where she and her husband are the caretakers of a large farm. Terry and her husband are in the process of moving closer to town (Tonasket, WA) but Terry took time out of her day and showed us her goats. Milk goats - to be precise.

After *years* of saying she "hates the stuff," my mom (Cheryl) tasted goat milk at Terry's farm.

Terry had milked earlier this morning and brought out a jar of fresh milk from her refrigerator. Mom was skeptical but she took a small sip. She made a face and said, "I really can't tell the difference. It tastes like cow milk!"

Amorette told me later that she'd wanted to take a sip also. I guess she'll get her chance soon enough!

Kyra (three-years-old at the time) was able to bottle feed a baby goat which Terry was raising for her own. Terry had warmed up the milk in a tub of hot water in the kitchen sink. We followed Terry out to the barn.

Kyra was given a bottle and we went into the small pen. Out from behind the heat lamp came two beautiful white goats. They were pretty small, maybe a week old. Kyra had a huge smile on her face.

The baby goats came right up to us, they were friendly, loving and curious. These two kids were being raised for new milkers, so Terry made sure she handled them daily.

Kyra did as directed by Terry and I kind of held the bottle underneath so it wouldn't be too heavy for Kyra to handle

The goat drank quite a bit before it got curious about the rest of us in the stall and walked away.

Kyra was so happy. Amorette took a picture of Kyra feeding the baby (above) so Ron would see the event. Daddy hates to miss out on these fun times.

An old dairy barn is set up for the goats. Terry has at least ten milkers that I can remember. And maybe six or eight kids. I was particularly interested in the Saanen, one of several breeds within the milk goat clan.

Her name is Katie and she has three kids. Every year she has three or four kids. Katie usually gives one gallon of milk per day for Terry. This is an excellent amount of milk! The goat was frightened by us at first. Katie had not been around small children like our Kyra, and we were strangers to her.

But after awhile Katie warmed up to us and stood by our side wanting to be petted and scratched. We thoroughly enjoyed her.

I must have asked Terry a million questions and she was *very polite* and tried to answer all of them. I wanted to make sure this was an adventure we were up to before we jump in with both feet. And it was. I couldn't seem to find one thing wrong with goats. And even more, I couldn't find any reason why we had waited this long to try them.

"Whether (pun intended!) you buy from me or somewhere else look over the goats you buy. Check their feet to make sure they aren't overgrown. Some growth is okay after winter but not excessive. And look at their back legs to make sure they are not bowed. This could stop the udder from growing normally." Terry Dean explains.

I was real happy when we left Terry's place. She seemed to care about her goats. Ron and I are often accused of spoiling our animals and I think Terry was happy to hear that. Terry wanted her goats to go to good homes. She had raised Katie from a baby and Katie was now six years old.

Our Milk Goats Arrive

Since we bought the goats in March we haven't had to winter with them yet. It was still pretty cold outside on our homestead in March, and we had a few snow flurries even the first week of April.

♥ Our barn was relatively draft free.

♥ We provided wood chips as a floor covering and made a nest of thick straw for bedding. The milkers seemed pretty happy and content.

♥ Ron and I made sure the goats had an electric heated water bucket to drink from. This way we would not have frozen water to break ice from.

♥ We keep the chores simple.

♥ We run a line of electric fence around the inside of our stock panels and this keeps the goats inside of the paddock. And it keeps the dogs or other predators out.

When it comes time for our does to kid I am thinking about adding a wood partition for the mother's, to keep more drafts out. It would be the saddest thing on earth to walk out one morning and have to peel a baby goat from the frozen ground!

Preventative measures are best in every aspect when you have animals to care for.

My New Milk Goat

Here, I will tell you our experience with the goats coming to our home. We keep a nice clean place here. The goats had no problem with that. They did miss their barn buddies from their previous home, though.

Katie came to our home first. Katie is a six-year-old Saanen goat. She is all white in color. It took Katie nearly two hours to arrive. I know she was stressed when she got here. Goats do not like to travel. Katie had two kids; one female and one male.

I didn't want to stress the goats anymore than they already were from traveling. We had the stall ready in the barn with clean chips covering the floor and a thick pile of straw for them to bed down in.

♥ It was still below freezing at night. The electric water bucket was filled. The hay rack was built and filled with good quality hay.

♥ I had decided to put the goats in the stall and try not to bother them too much the first couple of days. This way we could let them get used to us and vice versa.

The problem I saw right away were our dogs. Dogs are a natural predator for goats. Katie kept stomping her feet at our puppy, Jynx. She was very protective of her babies. Katie kept a watchful eye through the cracks between our rough cut two-by-six lumber which lines the barn walls. I knew she would head-butt either one of our dogs if given the chance, so I kept them away from the barn as much as possible.

With any new animal to your home you want to give the animal time to adjust to the new atmosphere; smells, barn buddies, and of course, their new routine. I would be milking Katie soon enough. For now I just want Katie to settle in and feel at home here!

Milking Once a Day

You don't have to milk your does twice a day as most folks feel is demanded. The goat will not bag up from milking her once a day as long as you follow a few simple rules. (See below for bagging up).

♥ If you keep the kids in a separate pen at night and milk your does first thing in the morning, then you can let the kids out with their mama all day and the doe won't need to be milked a second time.

♥ Keep to a constant schedule whatever you choose to do. Do not be sporadic about milking times.

Bagging Up

As long as you are consistent with the time of day chosen for this chore, the goat should not bag up.

♥ Bagging up tells the goat that her kids don't need the milk she is producing any longer and the doe's milk will begin to dry up.

Milking Katie

It has been one experience after another with goats! They are social and loving animals. When I go outside I can hear Katie calling to me. She enjoys being stroked and loved on. It has been nothing but a pleasure since Katie arrived.

At first I milked the goats twice a day, and I tried to keep to a 12 hour schedule with milking, but this was too much with all my other chores and I had to start milking once a day.

♥ I let Katie get used to her new surroundings for three days before I tried milking her.

♥ I was overly anxious to get started. I'm not sure Katie felt the same. She hadn't been milked this year, as Terry didn't want to bottle feed Katie's babies. But I wanted to milk a goat, any goat, and Katie was here!

♥ I separated Katie from her babies the evening before. You can't let the babies drink all of the milk or there won't be any left for you. Her bags were humongous the next morning. Katie must have been in pain with all that pressure on her udder.

♥ Katie leads well so I took her into the tack room and after a lot of coaxing she was up on the milk stanchion we had borrowed from a neighbor.

She wasn't used to our set-up though, and it took me quite awhile to get her to step up on the stanchion. Katie isn't a little girl so I couldn't just lift her up myself; she probably weighs one hundred and sixty pounds.

Eventually she stepped up and put her head through the opening. I pulled the rope over the two boards her head slipped through and we were ready for business. At least that's what I thought. Have I mentioned before that nothing is ever easy and simple around here?

Katie would not let her milk down. She seemed to be holding it in! I was at a loss of what to do. I had never milked a goat before and didn't know if I was doing it correctly. All I knew for sure was that there wasn't any milk coming out when I performed the milking art. *And it is an art.* Don't let anyone tell you, "It's easy."

Milking Goats: An Art

Katie's babies were crying out in the stall beyond the tack room. They were louder by the minute and Katie was getting restless listening to them. Her grain had run low. She was stepping back and forth. I was still trying to get a squirt of milk. All I wanted was one *single drop* of milk!

♥ I brushed her body to try and help her relax.

♥ Then I used the clippers to trim her hair hanging around her udder. Katie had four inches of hair all around her belly area.

I didn't get far in my chore before Katie was moving back and forth so much she stepped right off of the platform!

She struggled with her head which was still caught in between the two boards. I quickly undid the rope from the boards keeping her head there and she was fine. I gave up for the day and let Katie out with her babies. Katie would need more time to become aware of my expectations and her new routine.

Galen & Shannon's Goatly Crew

Our neighbors have been raising milk and meat goats for a couple of years now. I wasn't interested in goats when they *first* started their goat crew. But now I am. Shannon told me they want to travel a bit, which means condensing animals. I called to ask her if she would be selling any of her animals or equipment used for milking. She would be.

Shannon had two dairy does which would be perfect for my herd. I wanted to start a small flock and didn't want to get in over my head.

I asked Shannon to come and show me how to milk Katie. She was happy to oblige.

♥ Shannon was stunned at the size of Katie's teats. As she prepared to milk her she said Katie's bag was tight from being kept from the babies. Yes, that was true.

Shannon told me to watch while she milked Katie. And she showed me what is done in the pictures we studied in our books; the art of milking a goat!

See *Storey's Guide to Raising Dairy Goats* for illustrated guidance through this procedure.

Basic Steps to Milking a Goat

Step 1: Place thumb and forefinger together at top of first teat until you are clamping off the milk inside the teat. You do have to apply pressure.

Step 2: Gently squeeze thumb and forefinger together to release the milk from within the teat. Then without unclamping those first two fingers bring the middle finger around to push out more milk from inside the teat.

Step 3: Repeat step #2 with last two fingers or until teat is limp and empty. Start over.

Warning! Do not *pull down* on the teats during this process. It can tear the milk tissue within the teats.

Shannon has a few of her own milking goats and she was more than helpful and patient with me. I asked a *zillion* questions as I always do during the learning process. Shannon answered them all and more.

♥ When she got a big stream of milk from the teat I just about flipped my gourd! How did she do that?

Shannon was honest and told me that she had to 'learn' how to milk too, but she says, "Once you get the knack of it, it's like riding a bike." I bet she could milk anyone's goat!

Under Shannon's watchful eye I learned to milk Katie. And I have been milking her ever since. I love it. It's actually very peaceful and I can plan the rest of my day while doing this chore.

After my milking lesson, Shannon and I got to talking about her goats and Ron and I decided to buy two milkers and we have been very pleased with the results. And very busy!

After much practice, I am now able to teach others the art of milking a goat. Our daughters have friends who enjoy spending time on our farm. When they come to visit us we have lots for them to do and they actually enjoy this work. I guess it's because they're not at home so it doesn't feel like a chore.

Below is a picture of Gina Koenig. Gina is Amorette's best friend and while doing chores she was able to learn how to milk Katie.

Gina Koenig learns to milk

"It's easy," says Gina, "After you figure it out."

Did you know that the tail shows how your pig is feeling? A curly tail shows health and happiness. A straight tail shows sickness or distress.

9
Pig Basics

It is very rewarding and easy to raise a few pigs on a farm. We have enjoyed it immensely. The following basic steps to raising and caring for pigs can be the same if you breed your own or if you raise one pig a year to butcher.

A few words about the Pig

Barrow – Castrated male
Boar – Mature or immature male
Farrowing – Giving birth
Feeder Pig – Weaned piglet
Gilt – Immature female
Shoat or Piglet – Newborn
Sow – Mature female

Housing

For us, it makes the most sense to spend more money now to save us lots of money and time later.

Ron thinks like this during the planning stages hoping that he won't have to rebuild housing later if he does it right the first time. And the longer the housing will last, the better our money was spent. Makes sense to me....

I don't want to go into great detail on housing, since there are a thousand books covering that information. But here are the basics, and with the picture you can get a better idea of what we've done. It's worked for us for years!

Our 8 x 12 foot sow house

Approximate costs: Rosie's house cost us $450 to build and has held up fantastically the last four years. We expect this house to last another five to eight years. Not bad, right?

Here is a break down of the cost it took to build this pig house, depending how many years the house holds up to the weather and wear-and-tear of daily pig use. If it lasts us:

Four years: $112.50 a year
Eight years: $56.25 a year
Twelve years: $37.50 a year

♥ Our pig house consists of an 8 x 12 by three feet tall structure. Ron separates the house into two sides by dividing it down the middle with two by four boards.

♥ Add a door to the outside of each divide and Ta-da! Two houses and/or one large farrowing house.

NOTE! The doorway needs to be at least 40 inches tall if you want to allow for a growing pig. Our doorway is 36 inches tall and Rosie is starting to scrape her back on it now that she's four-years-old.

House on Skids

Ron uses treated lumber for the skids/runners. Using skids allows you to tow the housing to another location with ease. We use our tractor and chains to move them around.

We build all of our buildings and housing on skids now! It works great. Ron can build the housing down at the barn where his tools are easily stored, and we can tow it with our tractor to the spot we need the house.

Now we have a garden shed, pig house, and A-frame houses on skids for the butcher pigs. Ron is planning to build us another farrowing house soon too; and I'd bet money that it'll be on skids!

A-Frame Housing

Ron has built us many pens and houses for the animals here. It's wonderful to have a man that can build, and enjoy it while he does. Each house he builds is a bit better than the last, and he always seems happy with that.

For winter we use the farrowing house mentioned in the above housing section of this book. For spring we use A-frame houses on skids (so we can move them if necessary).

♥ In the spring pen there are two A-frame houses. Rosie and the other sows use one of the A-frames on their side of the electric fence pen. The grow-out-hogs that we raise for butcher use the other A-frame.

♥ We face the houses against any North winds. Shana and Amorette fill the dirt floor with straw to help keep the pigs warm if there is a chilly night. That's all there is to it!

Heat/Creeper for Piglets

When Rosie has her babies in February, the piglets will require heat in order to survive. We can put heat lamps on the other side of Rosie's divide and it will be safe from Rosie nosing it, knocking it down and starting a fire.

If nothing else, it's a good idea to have a small space where the sow can't get to the heat lamps because sows *are* curious animals.

Sheri Vejrostek has her story to tell here. One of her sows wondered what the heat lamp hanging above her was. The sow curiously pushed it around with her nose and when it fell the lamp got stuck to the sow, burning terrible second degree burns into her delicate pink skin!

Another farmer told me how he had a heat lamp in his dog house and the dog knocked it loose. The heat lamp came down and the dog house caught on fire!

♥ Heat lamps can be dangerous. That's why we like to have the divider; an entire separate area to hang heat lamps away from curious sows.

♥ Also, the piglets can get away from mama if needed and use the heat from the heat lamps for their source of warmth.

Added Note of Caution: Fires!

At Just-A-Little Ranch we had our own fire. When I woke up that morning, I stood by the stove to warm my toes and looked out at the pig house like I always do.

I saw smoke drifting out of the house like the pigs had their own chimney out there! It was one of the scariest sights I had seen in a long time. The smoke drifted out of the house at a fast steady rate and went spiraling up in a whirling loop into the air.

I screamed for the kids to wake up and help while I grabbed my shoes and ran out the door. The kids were right behind me. Luckily the fire was only smoldering, it hadn't caught yet.

♥ I chased the pigs out of the house (they were all sleeping). The piglets had moved away from the corner under the heat lamp where it was smoldering, and they were snuggling next to mom in the middle of the house. Not one pig was safe from the smoke.

♥ With the pigs out of the house I could see what had happened.

♥ The extension cord we'd used for five years had shorted out and sparked. The spark started the straw to smoldering.

♥ There was enough smoke above where I crouched to choke me. I began to remove the straw before it ignited completely.

Finally, with the help of Shana and Amorette, we got all of the smoldering straw out of the pig house and all was well. From now on we will make sure the extension cords are all in good working order.

NOTE! If the extension cords are kept outside year after year, I suggest replacing them every three years to be safe.

Fencing

On our homestead, we keep our pigs inside electric fencing. We use a solar panel to keep costs down, and the sun is all we need to run our electric fence for the pigs.

♥ The electric fence wire is run with two strands inserted into insulators which are attached to rebar posts every fifteen feet.

♥ The first strand of wire is eight-ten inches from the ground, and four inches above that one is the second wire.

In the spring the pigs we raise for butcher are inside this 150' by 150' pen. The wires can be lowered or raised higher by using the insulators that attach to posts.

Pigs Respect Electric

♥ You can keep an 800+ pound sow in an electric wire enclosure very simply. Just remember to teach the piglets about electric fencing while they are young; by three-four weeks of age.

Our sow doesn't even try to walk over the electric, even when it's set short for babies (its four and eight inches off of the ground). She was raised in electric fences!

♥ Remember, put up your electric wire for little ones right away. You will be amazed at how they can sense the fence is on.

♥ Once they get zapped a couple of times they won't go within a couple inches of it. They can smell it. They fear it.

♥ I even noticed that the grass line about four inches inside the fence doesn't get munched on. The pigs don't get that close to the fence once they get electrified!

Hog Panels

Hog panels are all we use to house our pigs for winter.

♥ If you buy hog panels, try to find the ones with the four rows of smaller squares at the ground level. The piglets cannot escape these.

♥ These hog panels are 32" tall (16' long), so they aren't as heavy as the 52" ones.

♥ You could run chicken wire along the bottom if you buy the ones that won't keep the babies in, but you might find cut noses!

If you can only find the hog panels with the big squares, not to worry, just run a line of electric fencing outside the panels about three inches off the ground and this should keep your piglets inside.

The piglets will put their noses out, get shocked, and learn respect for the fencing; and they can't get out into your yard to eat your flowers and root everywhere with their rock-solid noses.

Simple Feed or Water Troughs

We add holes in the bottom of our feed trough to let the rain drain out so the feed doesn't get all mushy.

♥ One hard plastic barrel cut down lengthwise will make two troughs! Then add a two by four lumber square structure, with ½ of the barrel built inside, and a metal t-post to pin it to the ground, and the feed trough can't be moved.

That is about as simple as you can get, and a whole lot cheaper to boot. Eating troughs for pigs can easily run you into hundreds of dollars.

♥ Pigs will eat off the ground of course, but when it's raining and the mud gets thick, why should they have to?

♥ We use these troughs to keep the feed waste to a minimum.

If you add two by two inch boards along the length of the structure it will allow only the piglets to enter the feed trough. This is a very cheap creep feeder! (Say that one five times fast!).

Free-Choice

I don't recommend free-choice feeding. I've had a few stories from fellow farmers who were not happy with the outcome.

Nick Baker bought a butcher hog at 240 lb.'s to slaughter out and eat. He told us there was so much fat on the hog he didn't get much meat and the pig didn't taste as good. Nick was very disappointed.

Nick then asked the seller what could have made such a difference in the meat since buying the year before. The seller admitted that all of his own pigs died this year so he bought replacements from another farmer. The other farmer feeds free-choice and the hogs got fat.

Did you know? Pigs only need to be fed a daily maintenance after their weight reaches 220 pounds. Any weight gain after this size is turned to *fat* considerably more than it is meat.

Greens

Give your garden scraps and all kitchen scraps to your pigs. This will offset their diet and give them more nutrition when they can't forage for themselves. It will also offset your feeding bill.

Try asking your grocery store for their garbage from the vegetable department. The store will trim back and throw out un-salable food which is perfectly grand for your pig!

If you can set up a deal with a local restaurant owner you can have a good off-set for those feeding costs. If you provide the container and pick it up when asked to, the restaurant will store the left over food scraps from day to day for you.

Cherries, Bread and Apples

Pigs are not picky eaters. If you raise hogs for butcher, then by all means, feed what you can get for free.

♥ But remember that pigs have a digestive system too, and it can get an ache if fed too much of the same feed.

At certain times of the year we get bins and bins of cherries and apples. We feed the pigs this feed twice a day, as much as they will eat in a twenty minute period.

It saves us a lot of money when we don't have to feed out so much grain. I give my pigs a couple of scoops of grain with the cherries or apples just to give them a better balance in their diet.

Grain and Hay

Ron and I buy bulk grain from our local feed store. The prices vary at different times of the year, depending if Omak or Tonasket Feed Supply can get the grain on sale.

♥ We load the grain into garbage cans which makes it easier for our kids to feed the pigs at home.

If you make the chores easy for the kids, you can be relatively sure they will perform the task. (You should always check their work periodically, though, they are still kids...) The amount of feed differs from pig to pig, depending on size and age.

♥ **Piglet from 30-50#'s:** Usually eat between four and six pounds of starter grain daily.

♥ **Pig from 50-200#'s:** Will need between six and ten pounds of grower grain daily.

♥ **Pig from 200#'s:** The pig now only needs a daily maintenance of six pounds of grain.

♥ **Test Your Pig's Appetite:** Let the pig eat all it can consume in a twenty minute period. The number of pounds eaten in this time frame is the amount you should feed out daily. Divide this between two feedings per day.

We toss the pigs all of our table scraps in addition to their daily grain.

Also, we toss adult pigs a flake of alfalfa hay every other day so they have something to chew on and they will receive added vitamins to keep them healthy.

Salt is Poisonous

Pigs are most often poisoned by eating excess salt in their feed which is sometimes accidentally added in large amounts. Or they can be poisoned by eating normal amounts on a hot day with very little or no fresh water available.

For more information about feeding your new piglet after bringing it home, see our documents in the resource section of this book, or visit http://www.foxmtnpublishing.com.

10
Piglets on the Farm

Mixing Sizes and Ages

It isn't wise to mix different sizes or ages of pigs. You might have a nice pig by itself but put it with another and you'll see a sudden change.

♥ If you have a lot of small pigs, keep them together.

♥ Keep the larger sizes of pigs penned together.

♥ Do not take piglets and put them in with older pigs or the older ones could eat them for lunch.

♥ If you buy a new pig and want to pen it with others its own size and age, first let them get acquainted through a fence line. After two weeks of this, try to let them into the same pen. They will have to establish who will be the herd boss but if they don't mortally wound one another it will be safe to leave them together.

Finding a Porker

If someone tells you horror stories of a pig that tore the flesh from some ones bones, it may not be true, but you *will* need to be careful not to purchase a mean sow or boar. If you have kids, you want them to be able to feed and water your animals without fear.

♥ When looking to buy a sow to start breeding, tell the owner of your intentions to breed. They may help you pick an exceptional female from the litter. Every pig has its own characteristics, and by the time we sell our feeder piglets I know which ones I would recommend for breeders.

♥ You can pick out your pig by its color and beauty first. Then watch it carefully. Reach out to pet it. It will come to you and smell your hand.

Keep in mind the piglets have to nip a few times in order to be taught not to bite. They are easily taught. Just give the piglet a smack on the nose and it should get the picture quickly. If a piglet doesn't get the point, don't keep that one.

♥ Watch carefully to make sure the piglet is not ill or carrying noticeable parasites. If the snout is dripping with wetness, aside from water that is, then the baby may have a cold.

♥ See that the piglets' eyes are clear and focused.

♥ Is the tail twirled into a curl? If it is hanging straight down then the pig is upset or ill. However, if you've just handled the piglet it could be stressed and have a strait tail.

♥ Piglets play hard and they nip each other in their play, so don't worry if you see bite marks on them or scabbed over spots, its just sibling love.

Try to see the piglet in its natural surroundings while it's not under stress. You don't want to start off your new adventure with a sick animal, so be careful!

And ask the owner lots of questions to make sure you know what you are getting into and what is expected of you. The only stupid question is the one that wasn't asked.

Handling a Piglet

I had never picked up a piglet, and I was sort of afraid but Sheri showed me how she does it. Of course, she'd been picking the pigs up for years and she made it look a lot easier than it is! But we tried. We just couldn't get those piglets to quit squealing. I was afraid one of us would go deaf if we had to hold them for too long.

♥ Sheri picks up the piglet by its belly and swings it over her shoulder.

She gets kicked during this process and that's another reason I wanted to find another way to pick the piglets up.

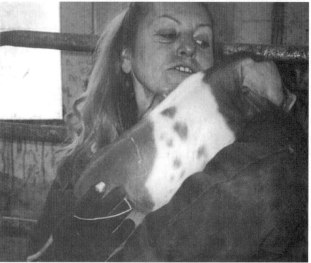

Sheri Vejrostek shows how to hold a piglet

Our first litter of piglets from our sow Rosie were ready to sell and people showed up to take their piglets home. What did we do now? Throw them over our shoulder? Well, no, because Rosie was upset to have us capturing her babies.

Ron decided to let Rosie out of the pen; we'd catch each piglet inside per pen and hand it over the fence to its new owner. Not the best idea. We knew Rosie was tame and would never hurt anyone, but she was upset that we were making those babies squeal. I mean shriek! They are *extremely* loud.

When Ron handed one piglet over the fence, the piglet was squealing so loud that Rosie came running from about 200 feet away. Fast!

People were hiding behind trees, jumping up on the hoods of cars, and bolting in every direction! It was funny to see (but only later when we thought about it). We caught piglets this way with Rosie's first litter.

Finally, during the Okanogan County Fair, a boy showed me how to pick up the pigs by their hind legs.

♥ Grab both of the piglets' legs with your hands, and hold them close together with the piglet hanging upside down.

♥ Piglets don't usually squeal when held upside down. They will kick and wiggle though, so be careful not to drop them!

During Rosie's second litter we caught all the babies by their hind legs, lifting them out of the pen without any fuss. We sold the piglets after the fair and Rosie didn't even know her babies were gone until her nipples reminded her that something was missing!

Traveling with Piglets

I brought a dog carrier to set inside the back hatch of our Jeep to take our piglet home in. It worked great.

I will tell you this: the pigs quiet down for the ride but they don't stop their bowels. And I am not talking about pee! You won't realize what I mean about the smell of pig crap until you have driven with it in a vehicle for an hour or more. It took us one hour and ten minutes to get home from Sheri's farm! We know the smell. Even a few years later...we remember the smell!

♥ A small or medium size dog carrier has worked best for us when traveling with piglets. It's safe, secure and easy to manage.

♥ Some people put piglets into a burlap sack; which is probably okay but I don't suggest it.

♥ I think you should try to keep the pigs as happy as you can during transportation to prevent them from stressing.

♥ You can put straw into your pickup (if it has a canopy) and let them run free until you get to your destination. This keeps them happy and secure.

Adjustment at a New Farm Yard

We took our new piglet (Rosie) home and let her out into her pen. Ron set the dog kennel on the ground inside her pen and opened the door. We waited for her to come out on her own. She sniffed around, looking at everything. She walked to her house, the feed trough, the waterer. Rosie even came over to us to sniff our hands.

I sat there with Rosie for hours in her house. I didn't want her to think she was alone, even though we had bought one of her brothers (for a butcher pig) to keep Rosie company.

I also wanted her to bond with us so we could breed her when she grew up.

For the first few days the piglets seemed to have diarrhea. I thought it was caused from the lettuce I had given them, but it was due to stress. Many piglets may go through this. They've just lost mama, and they're at a new farm yard. Some pigs are better adjusted than others of course. To make the change to a new home easiest, you should do the following:

♥ Gradually change the feed that the piglet has been eating over to your own brand. Doing this gradual change over one week's time is normal.

♥ Keep feed and water assessable at all times. Pigs can drink up to four gallons of water a day when full grown.

♥ Worm your new piglet right away. Wormy pigs don't grow well, they show pot bellies and their hair will be dull and stick out instead of lying flat.

♥ Take a fecal sample to your vet to check for parasites. Parasites cannot be seen with the naked eye; a telescope is necessary.

♥ Let little piglets get used to a new batch of piglets through a fence before putting them in together.

♥ Supply lots of fresh straw to burrow in.

♥ A draft free house is needed in winter.

NEVER!
♥ Never put new babies in with older or larger pigs.

♥ Never catch a pig by its tail. The tail is a part of the spinal system.

Teaching Piglets Respect

It is very simple to teach piglets to respect an electric fence. If you put the electric fence up when they are only a few weeks old they will respect it for the rest of their lives! We keep two eight hundred pound sows inside two strands of electric wire. Boars require different fencing, see chapter 11.

All of our piglets are put inside electric wire fencing by the age of four weeks.

Castration at the Vet: Piglet

This is the first time we took our piglets to the vet to be castrated. This is very expensive and we plan to learn to do the procedure ourselves from now on.

Luckily we only had three males in this litter of thirteen. The cost was $30.00 for the three. That seems pretty steep considering it only took the vet and his assistant about fifteen minutes to do the procedure.

It was quite the event to collect the three buggers and contain them in our trusty dog carrier. The first male we caught would squeeze out of the carrier while Shana was trying to catch another male. We only had the three to catch *but* we had to collect them from in between the other piglets. What a chore!

♥ I recommend you think it all the way through before you jump into a pen with the mama pig and try taking a few of her babies away. It's just not a good thing to make those babies squeal with mama right there.

It wasn't long before we had the three male piglets loaded in our dog carrier and were ready to go. I loaded them inside the back of our Jeep. We had added a nice amount of straw to the carrier so the piglets would stay warm. That's the problem, you see, you still have to keep those piglets warm (close to 80 degrees).

Since we live 45 minutes from town I made the appointment for Jynx (Shana's dog) to get her shots at the same time. What a mistake! We had to keep the Jeep as warm as any of us could stand for the piglets. All of us were starting to feel sick to our stomachs by the time we got to town.

Especially Jynx! The dog was drooling all over everything; the seat was thick with wetness, Amorette's coat was soaked and Kyra was starting to complain that "Jynx is getting my car seat all wet!"

Just before we got to town the dog threw up. Thank goodness it was on the floor. But poor Amorette and Kyra could smell the vomit the last 6 miles to the veterinarian, and that had to be bad with the sick feeling we all had *from the heater being on!* It just wasn't a great ride. That's all there was to it.

I told my friend, Debbie, I would bring the piglets for her to see while we were in town. I pulled up in front of Maverick's Bar and Grill in Tonasket, WA and the kids jumped out to get some fresh air. Who could blame them with the smell of dog barf and piglet defecation in the air?

Shana went to get Debbie. Debbie and Dan Perdew own Maverick's Bar and Grill. I will mention the *Blu-Baron* is the best sandwich I have eaten in my entire life. It is made on sourdough bread with blue cheese dressing, roast beef and swiss cheese. And the nachos are an enormous pile filled with all the goodies; the kids and I have leftovers when we share an order!

Dan & Debbie Perdew of Mavericks Bar & Grill, Tonasket, WA

"Oh, Jeanie, they're so cute!" Debbie said. "They're so little..."

"Yeah," I said. "They're *always* cute when they're *this* little!"

So, we get to the vet and Jynx leaves some presents on their nice perfectly green lawn. Shana and I bring the piglets inside so they won't get chilled.

I'm thinking, "Can I go home now?"

Kyra has to go potty. All is well. Cody Ames, D.V.M. finished up with the dog ahead of us and we took the piglets in for their procedure.

♥ His assistant caught a piglet from the dog carrier and held it firmly by the back feet, upside down.

♥ Cody then carved out two slits and dispersed the testicles on the table next to him. Wow. How would you like that done without any anesthetic?

The pigs were castrated without problems. Shana took Jynx in to get her shots and we were out of there. I just wanted to go home. The ride home was much the same though; *45 minutes of miserable heat and stench.*

For a complete set of directions to castrating piglets yourself, please refer to the end of this chapter.

Shana holds a newborn

Inbreeding Pigs

I don't suggest inbreeding pigs at all. It can be done, of course, if you know what you are doing. You should contact someone who is an authority in this subject. However, I am not taking any chances with my babies!

Care for Newborn Piglets

Ron and I don't believe we should inflict unnecessary pain, so we don't give iron shots, clip the wolf teeth, dock tails, or castrate our piglets. Let me explain!

Iron is Critical

Yes, the piglets need iron within three days! It is detrimental to their health. But we don't give the pigs a shot. We give them *dirt*. Yes!

I have one of the girls dig up a three gallon bucket full of dirt and dump it on the piglets' side of the divider in the house, and the piglets root and eat the dirt. It has worked for us through three litters of healthy babies so far!

♥ You see, all you need to do is add dirt from your ground on day two and day four, then let mama do the rest. She will teach her babies to root.

♥ You should pet and pick up your piglets so they get used to the human touch, too. We do this on a daily basis.

♥ Of course, if it's winter time you'll need to add dirt every other day until the babies learn to root on their own. The ground will be frozen and they will need help. Babies will follow mama outside when it's nicer and start rooting on their own.

Wolf Teeth

The mother sow will let her babies know when they are being too painful on her nipples. Rosie usually lies on her side to let her babies nurse. She will happily grunt every few seconds, most mamas do this.

But, if a piglet is nibbling or biting her nipple, Rosie will rise up and not let any piglets nurse for awhile. And you can tell Rosie is punishing them by the different grunt; it's actually sterner if you can imagine that!

If you still feel you would like to clip your piglets' wolf teeth, here is the way to proceed.

♥ There are eight wolf teeth in the pig; two on each side of the upper jaw, and two on each side of the lower jaw.

♥ These wolf teeth should be clipped off at the jaw line.

♥ For this task you can use small clippers or sidecutter pliers. Do this job away from the sow.

Dock the Tails?

What the hell for? That's what I want to know! If I wanted a Rottweiler, I'd buy one, right? We don't dock pigs' tails, it's just torturous! But here are the directions for you.

If you are keeping a boar or gilt to sell as breeders, remember that docking the tail is considered an undesirable characteristic by some people.

♥ Did you know that a curly tail tells you the pig is happy and healthy? If the tail is straight down then the pig is in distress or is sick.

♥ Don't clip the tail too short; it could lead to problems.

♥ Use sidecutter pliers or nippers to remove the last half or two-thirds of a pigs' tail at the same time you finish other newborn piglet care.

♥ As with all of your other piglet care jobs, perform this task away from the sow.

Castration

Now, we do castrate but we let Bill Vejrostek do the job. Ron can barely think about the process let alone actually try and castrate a piglet. He's too weak and so am I.

There is only one reason not to castrate a male piglet and that is if you want to keep the male for breeding.

Otherwise it is important to get the castration done right away. The longer you wait to castrate the male piglet, the harder it is on him.

♥ We castrate at two weeks of age. The male seems to heal quickly and forgets the whole incident a lot faster than when we castrated at five-weeks of age.

♥ To make the incision you'll need a disposable castration knife found in local feed stores. They cost less than one dollar and work great.

♥ The incisions are small and exact.

♥ Make a small cut over each testis and through the scrotum. Then draw the testis out and sever it along the supporting cord.

♥ Treat the surgical wound with an antiseptic or wound protector.

Beware of the Sow!

Do not perform any of these newborn tasks near the sow. A sow is likely to get very upset when her piglets start squealing. Remove the piglets from view and hearing distance of their mama.

Boars will lean against anything that will hold (or crunch under) their enormous weight

11
When to Keep a Boar

Bringing our Boy (Romper) Home

If you only have a couple of sows then it won't pay off to own a boar; renting one is more economical.

When we had two sows we had to monitor when they came into heat, then load them into a truck and drive them an hour away to be bred. With only the two sows this wasn't a problem. But our eventual goal is to own six sows with one boar to service them.

♥ We had to keep close tabs when we knew our sows' menstrual was on its way so when the sows vulva was swollen and they were definitely in heat we drove them over to be serviced by a boar.

135

Three days later we loaded them up and brought our sows back home.

Now that we have four sows it doesn't make much sense to go through all of that trouble. There would be a lot more paperwork to keep tabs on the heat cycles for four sows.

♥ Sows come into heat every 21-24 days. This would be a lot of work for someone as busy as me!

So, we decided to buy a baby boar and raise him up friendly and accustomed to people. We hoped to know his temperament by spending a lot of time with him.

Bill and Sheri Vejrostek adopted a female hog to their farm which wasn't related to any other hogs present. Our boar was reared from this hog, her name is Patches.

We named our boar Romper. The name suits him since it tells his job on our homestead (Romp-her). I planned to name all of our sows with names beginning with the letter 'R' just to be cute. There is no reason not to keep up the tradition with our boar.

♥ Romper was very well adjusted when he came to our homestead. Bill and Sheri get right down on the floor with their baby piglets and coexist with them.

Bill plays with piglets

♥ This lets the pigs know humans are all right and they aren't going to hurt them.

♥ Bill and Sheri spend hours with the piglets in their laps, petting them and picking them up so they get used to the human touch.

♥ Their philosophy being the piglets will be friendly and therefore adjust more readily to a new home when weaning time comes. And it's true!

Romper in the Barn

When Romper came here we made up a 12' by 12' stall in our barn for his house. Shana set up a heat lamp since it was still twenty degrees outside most nights.

The girls and I sat in the stall with Romper for half an hour that first night hoping we could help him adjust to his new home. He expressed nothing but happiness for our efforts.

I sat next to the wall with my legs stretched out flat in front of me. As I waited for Romper to wander over I watched as he familiarized himself with his new house.

♥ Romper wandered around the stall sniffing the water and feed dishes, tasting the grain, rooting through the straw bed and then he finally came over to me.

♥ He climbed right into my lap and actually laid his head down on my legs! I guess he was exhausted!

Amazed, I just sat there enjoying his attention. Romper wiggled on my lap but didn't dart off as I had expected him to when I petted him. I rubbed his back and Romper instantly rolled over so I could reach his belly. Yes! This was a truly friendly piglet and that's exactly why we bought him from Sheri's farm.

Shana and our new boar, Romper

Preparing Romper for Life

Even after a week with us Romper still has not changed his friendly ways. He just adds to our glory as he climbs in our laps and reaches up with his nose to plant a kiss on our lips! He is *fabulous*.

The fact that he is red in color with a white stripe just adds to our happiness that he will produce beautiful colored piglets

♥ Romper will have his tusks pulled soon, and I want his needle teeth clipped down flat. This will help in preventing any problems when he's larger.

♥ I plan to spend half an hour a day for the first few months with Romper petting him and playing until he's too big to play with. By the time he's eight hundred pounds we should know his temperament and be able to trust him on our homestead.

We will always be careful with Romper since you can never fully trust what an eight hundred pound boar can be capable of.

Differences in Needs

A boar has different needs than a sow does. Make sure the boar is in a secure and tight fence! When a boar gets loose there are a lot of things to worry about.

♥ Boars can become 1000 pounds or more, measure up to four feet tall, and get to be eight feet long!

♥ If loose, not only can a boar service females when he's not scheduled but the boar can damage lawns, gardens, vehicles, buildings and just about anything you have around.

♥ Boars are very curious in nature and like to rub themselves on anything that will hold (or crunch under) their weight!

Fencing for Boars

Ron and I decided while Romper is six weeks old we will teach him to respect an electric fence. When he grows too large for this type of enclosure (approx. 500 pounds for a boar) we will run a line of electric along the inside or outside of his animal stock panels (hog panels) for extra safety.

Preventative measures are always easier than corrections.

The Mature Boar

One boar can service up to ten sows!

♥ I don't recommend owning a boar until you own at least three sows.

It would be less expensive (if you only own a couple of sows) to take the sows to a rented boar to be bred.

Having more than One Boar

We do not intend to own more than one boar at a time. Romper will be the only boar with us until he becomes too old and has to be replaced. Then, we'll start with a six-week-old boar piglet again and when he grows old enough to service our sows we will replace Romper.

If you have more than one boar on your property you should be aware of a few things.

- ♥ Boars will challenge each other even if they are raised together. It is normal.

- ♥ A secure fence will keep boars apart. When loose they can do severe damage to each other. Sometimes the outcome is fatal.

- ♥ To break up two boars in the process of fighting, you will need to bring a board down between them.

- ♥ I would also suggest using an electric prod to separate fighting boars. If one boar is stopped immediately and taken back to its pen, the boar should forget about the other boar soon.

Feed

You can follow the feed scale guide in the basic pig care (chapter 9) for feeding your boar. Just remember, he will require more feed as he grows bigger, and bigger, and bigger...

Water

Keep clean, fresh water in front of the boar at all times. In the winter, if you can, provide a stock tank heater inside the water trough so the water doesn't freeze over.

- ♥ Pigs can drink up to four gallons of water a day.

Story: Dog runs down Boar

My husband told me a sad story about a German Shepherd who chased down one of his colleagues' boar.

First of all the German Shepard would run and chase the cows and bite at their hind quarters. The cows would kick at the dog, but the dog never stopped his nasty habit.

The owner of the German Shepard had an 800 pound boar grazing in the field, minding his own business. When the dog found the boar he started barking and chasing him. The boar ran as best as he could but couldn't move fast enough because of his weight and the dog bit his scrotum sac open! The poor boar was a mess.

The guy and his wife tried to hold the testicles together and coax this bleeding boar into a squeeze chute. The boar was castrated because there had been too much damage done to his scrotum sac to repair. It was very traumatic for the boar.

Prepare for Predators

♥ Introduce your own dogs to your animals at once and make sure the dog knows that it is not okay to attack or bite an animal.

♥ Neighboring dogs get loose; make sure your animals are safe from attacking predators.

12
Swine: Sow Pregnancy Care

Finding a Boar

In early October we put an ad in the Buy-N-Sell paper requesting the services of a boar for our sow Rosie. Some folks who live up Tunk Mountain near Tonasket, WA answered our ad. For a fee of $35 we could breed our sow to their boar. I couldn't feed a boar all year long for that price so it seemed more than fair to me.

It turned out that the boar came from Sheri Vejrostek's farm, which is our sow Rosie's original birth place. This news told me the boar would be friendly and treat my Rosie well. Sheri and Bill take great pride and a lot of hours with their pigs to make sure they are full of love and happiness.

♥ When you find a boar to breed with your sow, check to make sure he is in good health.

♥ Keep in mind that it will be a better match if the boar and sow are approximately the same size and weight. You don't want him to crush her!

♥ Pigs pass good or bad traits down to the piglets.

See that the boar is parasite free. We spray our sow with Zonk It before she goes to another farm just to keep any parasites off of her while she's visiting.

Flush the Sow

♥ I knew that Rosie should get seven to ten pounds of grain for seven to fourteen days to flush her system *before breeding.*

♥ This feeding schedule increases the ovulation process (egg production rate).

♥ By flushing a sows system you are increasing the number of eggs dropped during ovulation, which increases the number of piglets per litter.

As soon as we found a boar, I asked Shana and Amorette to start giving Rosie two extra scoops of feed to flush her system.

Introducing the Happy Couple

It's important to introduce the happy couple to each other through the safety of a fence line.

Let the sow in with the boar so she doesn't feel she has to protect her turf. This will be much easier on her.

If the sow is in standing heat she should fall directly in love with the boar. If not, break out the candles for a day or two and let the couple get acquainted.

Was she *really* Bred?

There are a couple of ways to tell if your sow has been bred or not.

♥ If you see the breeding dance yourself.

♥ For a first time breeder you will most likely see a bald spot on the gilts back. This bald spot is from the boars hooves and will be approximately two inches in diameter. This is what Rosie showed us after she was bred the first time.

♥ Older sows know what they are doing and it is harder to see any signs of their having been bred. Keep the sow in with the boar for two heat cycles to be sure of a breeding.

Rosie's second breeding was a visible sight to us, so we knew she'd been bred. The only clue to Rosie being busy during her third breeding visit (with Abraham) was her ruffled hair. It was sticking up all over the top of her back where the boar's hooves rested while breeding! I asked Rosie if she was going for the 'spiked' look but she wouldn't say.

Use Caution for 14 days after Breeding

After breeding the sow, take extra precautions not to let her run, jump, or have a harsh fall for ten to fourteen days.

- ♥ The eggs from ovulation are not cemented to the wall of her uterus until *at least* ten days after breeding. This is a critical time for a sow and accidents could shed unanchored eggs.

- ♥ If you loose eggs from her wall, then you've lost piglets. I didn't want to loose any eggs, so we put thick straw in the back of Big Red (our ugly farm truck) and drove home very slowly.

- ♥ If you can keep your sow in a separate pen for this time period (10-14 days) you can be pretty certain the number of piglets will be greater than if you didn't separate her. I've tested this theory and it works!

Gestation

Three months, three weeks and three days is the pigs' gestation period. Or, 114 days. We bred Rosie to Gordy on November 6. Her expected farrow date was February 28.

Ron and I wanted to make sure we did everything right so we invested in all of the pig books our local feed store carried. To our dismay, most books didn't go into great detail and left us with many questions about farrowing. I hope to answer all of your questions in the following chapters.

The first Three Months

The first three months the sow is pretty much the same as ever. Just feed her four-six pounds of grain for daily maintenance and make sure she has all the fresh, clean water she needs. A sow can drink up to four pounds of water a day.

Last Month of Gestation

When the sow is in her last three weeks of gestation all the rules change! The last month is when the babies make the most demands on the sow's body.

♥ The sow will need more feed at this point. Add two pounds to her current daily ration.

♥ If you have it, allow her a leaf of alfalfa hay every other day so she is sure to receive all of the vitamins she needs during pregnancy.

♥ Spend quality time with your sow now and she will be more apt to let you help with farrowing when the time comes. Brushing her down daily is a good way to start gaining her friendship.

We wanted to be there when the time came for Rosie to have her litter of piglets. The girls started giving her more grain rations the last three weeks before her due date. Rosie was approximately 800 pounds by now, and we fed her seven and eight pounds of sow chow daily.

Second-Belly Stage

Sometime in this three week count down, the sows belly will drop. It literally looks like she has a second belly below the first one! She will be *huge!* This is when you positively know your sow is going to have piglets!

If her belly hasn't dropped by the last three weeks before her expected due date, you can bet money she isn't pregnant.

2-Weeks Before Due Date

The sow will appreciate some extra attention at this stage in her pregnancy. You know how testy and emotional these females get... try giving her some juicy apple slices and a scratch behind the ears.

And remember to talk to her. If she has problems during farrowing, you'll need to help, and it will be much easier if you have spent the time gaining her trust and respect.

♥ It is a good idea to worm your sow now so her body is flushed clean for farrowing.

♥ Check with your veterinarian or local feed store to see what worming medicine is used in your area.

Two weeks before the due date we give our sows some worm medicine in pulverized grain. We haven't had problems with worms on our homestead but all the books said to worm her so we did. I checked our sows' droppings daily after worming and there were no worms visible.

1-Week Before Due Date

♥ Now it's time to prepare for the piglets. If you haven't already, get your heat lamps in place if it's below 50 degrees outside.

♥ The nipples should have dropped into v-shapes and the lower belly should be full of milk.

Nipples drop into v-shapes full of milk

♥ Soon you will be able to squeeze milk from a nipple.

Signs of Labor

You will have a few signs that your sow is in early labor.

♥ Her breathing will become ragged and she will begin to pant more often as her time draws near.

♥ She will get up and down in an attempt to become more comfortable. This early stage can last eight to twelve hours.

Hard Labor Begins

The sow will not move around as much at this point. She will spend most of her time lying down, and her breathing will be more audible.

♥ Eventually her body will shake a bit, her tail will twitch and she will seem to push now and then.

♥ The sow may expel a bit of blood or clear jell-like fluid. This is normal.

♥ This stage could last another 4-8 hours, depending on the sow and her experience.

Sow while Farrowing

Every sow is different but you will learn your sow's ways. See "Every Sow is Different" at the end of this chapter for more comparisons.

♥ The sows' tail may twitch just before she farrows each piglet.

♥ Pigs cannot stretch their heads to see behind them so she may stand to see what has happened back there after having a piglet.

♥ Make sure the sow doesn't step on the baby if she does stand up after each baby enters our world.

♥ It is normal for piglets to be born head or feet first.

- ♥ Piglets are covered with a wet film barrier they have to break out of to survive. It is best not to help them with this act.

- ♥ Babies are born with the instinct to nurse and they quickly find their way to the line up.

Human Intervention

Pigs have been giving birth for thousands of years without our help.

- ♥ If you do need to help a piglet into this world, make sure your hand and arm are sterilized. You can give the sow an infection if you're not clean.

Don't Forget about the Sow

Now that the sow has done her job don't forget to give her a pat on the head and a treat for a job well done. Some excess lettuce from dinner is good.

- ♥ The sow will want a sip of water. Get a bowl and put it to her mouth and she will drink it.

See my document #225 for feeding the nursing sow and taking care of the newborn piglets. Check our resource section in the back of this book; also see our website at http://www.foxmtnpublishing.com for more documents.

Sow feeds her piglets after farrowing

Enjoy!

Now is the time to enjoy the piglets for they grow quickly and will be gone before you know it.

Every Sow is Different

Each Sow Farrow's Differently

In this area I will list my two sows and a few other farmers' sows. My reason for this is so people can read, compare and realize that every sow is different, just as every woman is different when giving birth. It is much the same for pigs.

152

However, a few common facts do occur in every sow. For example:

♥ The dropping of the milk bag. It may drop three weeks before the sow is due or it could be one week prior, it depends on the sow. Yet it is unmistakable when it occurs.

♥ The sow will develop what looks like a second belly below her actually belly; this happens to be her milk coming in.

♥ There is the unusual sow which will not develop any milk. This is rare, though.

♥ When the week before farrowing arrives, the teats will slowly fill with milk and become very tight and full.

♥ The first milk coming into the teats contain what is called colostrum. It is important for the babies to get this colostrum. Piglets will have a much stronger survival rate with the mother's first milk.

♥ A day or two before the farrowing event takes place the sow will become restless. She may come in and out of her house. She may look behind her more often; sniffing the air. It depends on the sow. Mine have skipped this trait as they grew older.

♥ Eventually the milk bags will be like a rock and you will be able to squeeze the colostrum from them. This may be uncomfortable for the sow, however, so be nice!

153

- ♥ When your sow starts collecting bedding material and making her nest, you know farrowing is imminent.

- ♥ There could be a difference in the sows breathing up to 24 hours prior to farrowing as well. She will pant more loudly as the pain increases and the time draws near.

- ♥ A day prior to farrowing some sows 'wink' at you with their vulva. This means that their vulva has a pulsing look to it, like she is tightening and un-tightening her vulva muscles.

- ♥ And most sows lose their appetite, although none of mine had.

Our Rosie's colostrum came in twelve hours before farrowing her first litter. Eight hours before her second litter. And Rosie didn't show any sign of milk in the nipples before her third litter! I tested her teats for milk at four o'clock with no results and she farrowed at eight o'clock that night. Every sow is different, as is each litter.

Rosie made her nest directly before farrowing. Now she creates it *as* she is farrowing. Between each piglet she tosses the straw around.

- ♥ If you watch your sows you will learn their behavior but remember they may be completely different next year. Mine are.

Sheri Vejrostek has a farm with many sows which are bred every year. She has let me use a few of her sows' pregnancy diary's to give you a better idea of what a sow goes through while farrowing.

I will add Rosie's farrowing history in a completely different chapter for your knowledge and comparison. She has had three litters with us and I have documented them all. I will put the events of two pregnancies in order for your comparison. (Litter's number one and two are also in my first book; *PIGS; and other stories*). See resources.

Sheri's Sow; Patches

Patches is the mama of our new boar. He is ½ Yorkshire and ½ Landrace. We named him Romper (Romp-her). I felt that was cute and it fits his purpose here!

Patches is a new sow to Sheri's farm. Sheri took her and two others in when the owners could no longer feed them. She didn't realize Patches was pregnant until three months later when Patches started showing.

Patches is of the Landrace breed. Bill and Sheri didn't know her temperament since she was new to their farm. Sheri decided to bring Patches into their 'hog work room' next to their home so she could spend one-on-one time with her and gain the sows respect.

♥ Every day Sheri would brush Patches with a dog brush and get acquainted with her. Sheri is dedicated to her animals.

♥ By spending the week with Patches before she farrowed, Sheri was able to see what a great personality and docile temperament Patches

155

possessed. This would come in handy if Sheri needed to help the sow during farrowing.

Patches did a wonderful job giving birth to ten perfect piglets. The first piglet was born at 10:00 P.M. and the last was squeezing out at 3:00 A.M. Five hours of labor is pretty common among sows.

Sheri was present during each piglet's arrival into this world. She picked each one up, checked its health and set it under the heat lamp. There were no medical interventions necessary.

Sow: Babe V

Also from Sheri's farm, this sow's name is Babe and they add a V after this to distinguish her between another pig with the same name. We are going to follow three days prior to farrowing to see how this sow acted. Here's the diary.

February 18. Seems to stretch back legs out. Very little dilation. Lying around more today on side. By feeling the mother-to-be's stomach; the babies are more active today. Babe V is twitching her tail. Her bottom (vulva) is dark pink just as her teats are.

February 20. Some dilation. Babies quiet last night. Sow is peeing on different areas in bedding.

February 21. No appetite. Babe V is making a bed (nest). She is nervous and seems to re-make the bed. Restless! 1:07 P.M. Sow is dripping clear jell out of her vulva. 1:27 P.M. No babies yet.

Pushing a lot. 1:28 P.M. First baby is breech, pink. 1:29 P.M. Second piglet is pink. Third baby is pink.

156

2:51 P.M. Fourth baby is breech. Had to help the piglet out of the birth canal. One leg came out, then Sheri pulled gently and the second back leg came out. 3:35 P.M. Fifth baby is a pink female.

4:02 P.M. Babe V discharged a piglet from her vulva which was wrapped up in afterbirth too thick for the piglet to break out of. Sheri helped this piglet out of the afterbirth and the little female seems fine.

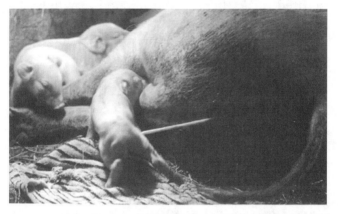

Umbilical cord is still connected to piglet

4:12 P.M. Baby number seven is a pink female. The umbilical cord didn't release out of the vulva as it usually does. Piglet is still attached to mama yet trying desperately to get around the corner to a teat!

4:18 P.M. Sow raised up and turned around. 4:19 P.M. Baby number eight is born. 4:42 P.M. Breech. Sheri helped baby enter this world. 4:56 Tenth piglet is another breech with only one leg emerging from vulva on its own. Sheri helped pull the baby out as the sow gave another push.

Our #2 Sow: Puller

Puller had her first litter of piglets on our homestead three weeks prior to our sow Rosie. I wasn't sure what Puller would do if I had to stick my arm inside her vulva to check how the progress of her labor was going. She seemed docile and warm to us but we couldn't be sure yet. She'd been on our property for a meager ten weeks.

♥ I wanted to be there if she needed any help. I spent a few extra minutes each day talking to her, petting her and trying to get acquainted. Puller seemed to like me just fine.

♥ Puller's bag of milk dropped three weeks before she actually had her babies.

♥ Our other sow doesn't show like this (showing the second belly stage) until a week before she is due. I thought Puller might be closer to having her piglets than we all thought but I was wrong. She held out for the three weeks until she was due to farrow.

After her belly dropped and I noticed a caved in area in front of her hips, I started giving her sow-chow and twice the feed ration. Puller didn't disperse any milk when a nipple was squeezed. I tested for milk daily.

She never made her nest until she had her babies. And we found out she had her babies *after* the fact. Usually I like to be there while my sows have their litters just in case they need help and to bond with them. Puller gave us no indication of imminent birth. I truly felt gypped of the experience I missed.

♥ Puller gave birth to fourteen beautiful piglets while we all slept!

I guess she didn't need my help! One baby was stillborn and we removed it from the house as we scooped up the afterbirth for the burn barrel.

All the piglets looked healthy. They were squirming and squiggling all over each other under the heat lamp. I waited to make sure they all lined up to her nipples as she fed. Thirteen babies got the necessary first milk they needed. All was well.

Do not let your sows eat their piglet's afterbirth!

13
Swine: Rosie's Pregnancy Diary

In an effort to help you understand that all sows are different and each one can farrow differently *each time* she has a litter, I kept a diary of our #1 sow's three litters. I only recorded two here though, in an effort to keep you from getting bored!

Litter #1: Time is Close Now...

One week before the big date, we could see a difference in Rosie's nipples. They had dropped into long v-shapes (see photo on page 149) and were carrying a bit of milk. Within one day of farrowing I could squeeze milk from Rosie's nipples.

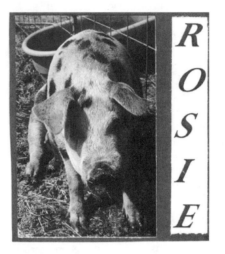

Signs of Labor

Finally, Rosie started signs of labor. At first it was mild. Rosie was breathing a bit ragged. This lasted for probably eight hours or so.

I spent most of those hours with Rosie, hoping my presence would soothe her. I would rub her belly and scratch her between the ears. I tried to explain how I had went through labor too, but Rosie didn't seem very interested, so I went inside to eat dinner and visit with the family.

♥ Rosie would rise up and say hi when I came to her pen but it was quite an effort for her to stand.

♥ Then Rosie started panting. She would sit up, pant, stand, pant, turn around, pant, and lay down. Pant, pant, pant. Again and again. Panting harder and seeming to push every once in awhile. It was about 9:00 P.M.

I took a sleeping bag, pillow, video cam corder and a bucket of water with clean cloths to Rosie's winter pig house. And a notebook to record the time of each birth. I wanted to describe each piglet as it came into our world so we would later know who arrived first and last.

Hard Labor Begins

♥ During the night her labor seemed to intensify. Rosie would lie in her house on top of her straw bed. Her body was shaking slightly and she was breathing quite hard. I was worried.

♥ By 6:00 A.M. it seemed the only progress she had made all night was to become more uncomfortable!

Ron made us breakfast at 8:00 A.M. We talked about our long, tiresome night. I don't think I got any sleep at all.

I continued to check on Rosie each hour. Her status remained the same. When she expelled a small amount of blood, I called Sheri right away.

"Since Rosie isn't due to farrow until tomorrow," Sheri explained, "I wouldn't worry, yet. It sounds like everything is moving right along."

Patience was not in my vocabulary, though. I was sure something must be wrong. At noon, Ron came down to Rosie's pen to visit and check on Rosie's progress.

♥ We all decided to come up to the house and make sandwiches for lunch. Before the cold cuts hit the bread, Pepper (our dog) was barking like crazy.

I looked out the window and Pepper was barking directly in front of the pig pen. We all hurried down to check on Rosie.

Surprise...

In the doorway, a small white piglet was wiggling its way out into the snow! Rosie had had three babies in the half hour we had left her!

- ♥ There was blood on the back wall of Rosie's house. I'm sure she had stood up and turned around after giving birth to look at what had happened back there, and rubbed her bloody behind on the wall in the process.

- ♥ Pigs cannot turn their heads around to see behind them. She had all these things going on back there and couldn't see any of it.

Slowly, so I wouldn't upset Rosie, I scooped up the loose piglets and put them into the heated box behind our divider inside her house.

Her house is eight by twelve. We had six feet to ourselves behind two by four boards for protection in case our sow got wild, mean or scared while we were in there. This way, I was with her while she was farrowing but if I felt threatened at any point, I could crawl over the barrier to safety.

Who knows what a first-time farrowing pig will do? We didn't. Better safe than sorry! It turned out the sow didn't do a thing. Rosie was the perfect hog in labor.

I sat beside Rosie while most of her nine babies came into our world.

♥ Piglets were born at odd intervals, some seven minutes apart, others nearly an hour. As each one burst from her vulva, Rosie would carefully (with the most care for a 350 pound hog) stand up, turn around and look down at her baby.

It was comical to see her take those little baby steps so as not to hurt the others while she looked at her newest baby. Then her labor pains took over and she would lie down again.

While she was straining with the moment of a new birth, I would scoop up the last piglet (or two).

♥ I wiped each piglet with a towel and put them into the heated box. I didn't want Rosie to step on them accidentally in her confusion or pain.

♥ The box was on our side of the house though and Rosie did grunt a bit about that.

Baby number seven was stillborn. The piglet took forty seven minutes to arrive. All I could assume was that the piglet had drowned in the birth canal. The piglet was large and perfectly formed. I tried to dry it off and rub it back to life but it never took a single breath.

Chart of Rosie Giving Birth

2/27/00		
	12:32	Pepper barking. Went out to see what for and Rosie had three piglets stumbling around her house. 2-males, 1-female.
	12:47	Baby #4, male, head first. Put two babies in box so Rosie can't accidentally step on them. Turned heat lamp on.
	12:53	Baby #5, male, feet first. Rosie is breathing ragged and hard.
	1:05	Baby #6, female, head first. Put two more babies in heated box.
	1:52	Baby #7, stillborn. Head first. Rosie was very upset after this birth. Seemed to be looking around for lost baby.
	2:04	Baby #8, female, head first. Rosie's breathing hard, visibly shaking. Some afterbirth came out with the baby.
	2:12	Baby #9, male, feet first.

As you can see by comparing this chart with Rosie's second litter (page 172), every pregnancy is different for pigs - as well as for humans. She was a bit overweight in her second litter, which will slow down progress and sometimes be the 'cause of problems.

Let Mom and Babies get Acquainted

Rosie had nine babies, eight which lived. Five white babies and three white with red and black spots.

Over the first couple of days, I was a video queen. My family didn't see much of me. I love pigs you see, and these were the sweetest kind -*babies!* Have you ever seen a baby pig? I went crazy off and on for two weeks and we made a video of how Rosie nurses, teaches the babies where to potty, games they play, and how vigorously they grow. Email me if you would like to buy a copy of my farrowing video Email: Jeanie@foxmtnpublishing.com.

♥ Rosie did not bark at her babies for nursing too roughly.

♥ Rosie carefully lowered down and nursed her piglets every couple of hours. If a baby bit her she would stop nursing, roll onto her stomach and not let any of the piglets nurse for awhile.

♥ Rosie's manners were impeccable with her little ones. Each one of her babies had its own characteristics and personality.

♥ By two days old the piglets would follow Rosie out into the snow and around the corner to her potty area! Rosie potty trained her piglets!

It's amazing how fast the piglets grow. When mama grunts babies come running. You can see the communication between mama and baby from day one. See our website www.foxmtnpublishing.com for documents about the care of your newborn piglets.

Litter #2: The Magical Day

It all happens when the sow is ready. No sooner. No later. I had been waiting a few days for Rosie to have her crew of piglets. As a matter of fact, I had been awake most of the two nights prior to her farrowing.

I had bought a baby monitor at a garage sale during summer to put in my sows' house. This way I would know at the click of a switch how her labor was coming, if there were problems and of course, when the babies started coming I would know and run to be with my sow.

On August 15th I took the baby monitor with me to the barn while I did chores, so I could hear if there were any changes with Rosie.

- ♥ She was breathing hard for the past two days, now the breaths were getting ragged.

- ♥ I noticed Rosie had made a nest in her straw. She was restless.

- ♥ There seemed to be a hollow looking area just above her belly in front of the hips (this meant her belly had dropped!).

I wasn't exactly awake at this point either. Like I mentioned before, I didn't know if Rosie would have her babies the 14th or 15th. I had been up off and on for the past two nights, so, of course, now she would start farrowing!

I decided I would just go visit her and see how she was doing. I took my notepad with me so I could write down notes. I brought the cam corder down too; just in case.

Rosie Farrows

I was inside her house for ten minutes, writing down some notes and getting as comfortable as possible on the other side of her divider when she started licking the wall.

"What are you doing, Rosie?" I said.

She didn't answer; just kept licking the wall with her back to me. That's when I noticed her vulva was wet. I looked closer at the wall and it looked like Rosie had smeared the wall with her vulva, and then licked it. Gross, huh? I think the wetness was the first broken baby sac.

♥ Not five minutes later Rosie laid down and squeezed out her first piglet. Wow. Piglets can be farrowed head first or feet first. Both are normal.

♥ Rosie's tail would start twitching and her vulva would swell more and more, then, all of a sudden, there was a baby easing its way out into the world!

♥ Piglets slide out of the birth canal fast. Rosie's tail stops twitching and she just lays there. The baby will cough a bit, jerk a little while it finds its ground and then it starts crawling to find a nipple to nurse on. Piglets are born practically running on their feet!

Identical to her first litter of piglets, Rosie would stand up after each piglet was born, turn around and check it out.

I collected most of her babies until she was done. I kept at least one baby with her at all times so she didn't get upset. Upsetting the sow can slow down farrowing.

I stayed with Rosie the entire time she had her eight babies. Four hours. I thought she was done farrowing because the afterbirth was emerging. I was hot and miserable. I had been inside her 8 x 12 house for four hours and it was eighty degrees outside.

I couldn't imagine how Rosie must feel but I thought she must want to be alone with her new family.

So, first I gave her a drink of water from the hose nozzle and then I let her babies into the pen with her. Rosie is so cute when she drinks from a hose.

♥ I dragged in the hose and turned the nozzle so she could drink some cool water. Rosie slurped it up and grunted a thank you to me.

♥ I left her to be a new mama. Her babies were playing and running and trying to find nipples, it was amazing what vigor they had for being hours old!

I rounded up my own three kids and we went out for dinner in Wauconda to celebrate. We had eight healthy babies. Not bad, Rosie!

Don't Forget the Afterbirth

♥ When we got home two hours later, I took a three gallon bucket and went to collect the afterbirth.

♥ The bucket was full and there was still *more* afterbirth oozing from her vulva! I went to dump the first bucket into the burn barrel.

♥ You don't want to let your sow eat the afterbirth. The sow might get the taste of newborn piglet in her mouth. This could cause a sow to eat piglets when she normally wouldn't.

When I came to collect the rest of the afterbirth I noticed a lot blacker and no belly showing when I looked at Rosie. She had piglets lined up all the way across her belly! I counted twelve piglets now!

Rosie had black babies, red babies, black with white, red with white, and one red with black spots! That was just one more beautiful thing about Rosie. Her colors kept you wondering what her babies would look like. She produced the prettiest babies I've ever seen!

Rosie and her children out in their yard

Next is an outline of Rosie's farrowing progress. This will give you more data of what to expect with your sow, although every sow is different.

8/13/00		No signs. Breathing: shortness of breath from heat.
8/14/00	11:00	Breathing: shortness of breath. Tummy has dropped. Milk is com-in in.
	1:00	Laid with her and rubbed her tummy. Could feel babies move and kick.
8/15/00	6:30	Breathing hard, not labored, not painful sounding.
	8:00	Breakfast; 6 pounds sow chow.
	9:30	Small grunts with fast breaths. Tummy tight. In labor!
	11:00	Rubbed her tummy. Breathing is labored, intermittent grunts. Small pushes (holds her breath and strains for split second). Went outside to lay in shade. Wetness on vulva.
	11:15	Licking wetness from wall; apparently from vulva. Possible baby sac?
	11:23	Baby #1: Black and white male, head first. Rosie gets up, turns around and looks at baby after each birth, but I will only list it here with the first baby.
	12:18	Baby #2: Red and white female, feet first.
	12:31	Baby #3: Black male, head first.
	12:43	Baby #4: Black male, head first.
	12:47	Baby #5: Red with black spots, male, feet first.
	12:57	Baby #6: Red with white on neck, male, feet first.

1:33 Baby #7: Black male, head first. Some brown afterbirth. Rosie is panting from heat, she's talking to her babies in box and two with her are nursing.

2:33 Baby #8: Black female, feet first. Afterbirth slowly coming out.

2:50 Short breaths. Afterbirth is oozing out slow as a seven day itch. Going to dinner with kids. Will let babies out with Rosie now. Gave Rosie drink of water from hose.

6:00 Came to collect afterbirth. Two 3# buckets full. Babies nursing all across her belly. Turned heat lamp on. Counted babies and found there are 4 more! Rosie had 12 babies! Two more solid blacks and two more black and whites.

Piglet Care is Easy

Piglets don't require much care. They need their mama most of all. Heat is the second most important factor. I collect each baby as it enters our world and put it in a box under a heat lamp. When Rosie is done farrowing I turn them all loose on her.

I did keep at least one baby with Rosie during farrowing though; it seems she gets nervous and upset if I take *all* of her babies away. But I didn't want her to step on any so I placed most of them underneath the heat lamp until she was done farrowing. See chapter 10 for more details about the care for newborn piglets.

A nursing cow can drink 35 gallons of water a day. Make sure it is available.

14
Cow Basics

As I stated in my book *'Pigs; and other stories'* we love animals. Ron, the girls and I like to give different animals a try once or twice to see how it goes. Its best if the outcome is profitable and the animals are holding their own. But most important is the happiness and the well being of the animals.

Raising our first cow, Bailey, has been wonderful. We were fortunate enough to find her at the auction sale as a day-old calf. She has been a pleasure in our lives ever since. The girls used to take naps with Bailey in the barn when she was little. All of us taught her to wear a halter and lead. Bailey follows us around like the dogs do. Our vet performed artificial insemination

(AI) in December and we expect Bailey to have her first calf in September!

Amorette kisses Bailey

If Bailey has a female she will become our second breeder. Our farm will profit if Bailey calves a male also, for he will be raised for the freezer. We win either way.

A Few Words about Cows
Bull – Immature or mature male
Calf – Newborn
Cow – Mature female
Heifer – Immature female
Steer – Castrated male

Shelter/Barn

A barn would be ideal, of course. But a barn isn't necessary for a cows needs.

♥ A three sided shed facing south will protect the cow/calf from rain, wind and snow adequately for shelter year round.

♥ In the winter it would be nice to add 6 inches of straw to give the cows an extra layer of warmth.

Fence

Fencing should be your first priority if you own a cow or are planning to buy one. Without a good solid fence, your cow will ditch you every chance he gets!

Remember how much a cow weighs, then add the fact that a scared cow will run right through a fence to get away from its predator, and you can see you need to put some serious thought into the fence, right?

There are many kinds of fencing available. I'll list a few but you should research and decide which would be best for your farming needs.

♥ **Barbed Wire:** Barbed wire works great if it is stretched tight and there are no large spaces between the wires. But this can injure your cow should it try to escape.

♥ **Barb-less Wire:** Is the same as barbed wire, only a bit safer for your animal.

- **Electric:** Electric wire is cheaper fencing and should be used more often. Cows can be trained to respect electric wire in a small area. Then you can electrify the rest of your pasture and keep the cow inside without many hassles.

- **Caution:** Do not run your electric wire parallel to a power line or telephone pole. If there is an electrical storm, it could cause a fire in the transformer.

- **Wood:** This is expensive and cows are notorious for escaping wood fencing. If it is all you have, run a line of electric wire along the inside about nose high to the cow and teach the cow to respect the electric wire.

Feeding Cows

We do things as basic as we can on our homestead. Cows are easy to care for but you have to remember that they have four stomachs and this requires a nutritional balance.

- A ½ an acre of nice green pasture is all a cow needs. This ½ an acre will provide adequate protein, vitamins, minerals and energy. Make sure they have water and a salt lick.

Ron and I have 10 acres of pasture that Bailey has access to with unlimited water and a salt block for three seasons; late spring, summer and fall until the grass dies. Then during winter months we toss her a couple sections of good quality alfalfa hay twice a day.

- During extremely cold weather provide a scoop of four pounds of grain to help the cow hold in her body heat.

- For one two-year-old cow we use about two tons of alfalfa hay per year and approximately 50 pounds of grain.

Poisonous! Hemlock and larkspur are poisonous to cows. Hemlock is also toxic to humans. Do not try to pick a hemlock plant.

If you want more in depth details about the cows' digestive system and extensive feeding plans, see *Storey's Guide to Raising Beef,* in the resource section.

Water

It's amazing how much water a cow can drink. Make sure your cow has a good size water trough, and a stock tank heater would make winter chores easier.

- A growing heifer can drink 12 gallons of water a day.

- But a nursing cow can drink 35 gallons a day

Squeeze Chute

A cow will need a squeeze chute to be given annual vaccinations or any veterinary care.

A cow squeeze chute can be as simple as you wish. The main idea is to trap the cow, tightly, inside an enclosure so the veterinarian can work on the cow with ease.

♥ Take a 10' stock panel and attach it to a wall with lag bolts that work like hinge pins. Keep the lag bolts in place until you need the squeeze chute put together. This way you can store the panel until you need it.

Worming

Worming is beyond the scope of this book, but it is important to set up a bi-annual schedule with your local veterinarian to worm your cows. Ask your veterinarian which products to use for your area.

Vaccines

Once a year you will want to vaccinate your cows. The vaccination schedules vary from state to state, however, just like worming medicines, so you will need to talk to your local vet to see what is used in your area.

Our vet comes out once a year and gives our cow a three-way vaccine which covers her for many different diseases which could be found in our area.

15
One Cows Artificial Insemination

I'm not sure if it's right to add our artificial insemination experience in this area of my book, but I'm going to be brave and do it. I want those who are thinking about using AI for their animals to know of the process necessary to accomplish the task.

Our cow Bailey has now been through the AI process and we felt it was relatively simple. I hope you will find this informational as well as humorous. And I hope more than ever to answer questions you may have about AI. The information I give will not be technical. I am not an authority on artificial insemination.

Preparing for AI...

We love babies here on our homestead. Without babies we wouldn't have much fun. All there seems to be to do around here is chores. And chores aren't always fun. If you have a baby to look at, pet and talk to while doing chores, the time goes by much faster!

Bailey was bought at the auction as a day-old calf. We named her Bailey because she is the color of Bailey's and Cream. Bailey is precious to us. Shana and Amorette spent many winter mornings napping in the stall with Bailey.

One time when Bailey woke up before the kids did, Shana awakened with a big surprise slurping her chin! Bailey decided Shana's chin looked like a bottle of milk and started nursing on it. We teased Shana for a week that she must have taken the vacuum cleaner and attached it to the end of her chin to get that rug-burn mark.

Our entire family spent lots of time training Bailey to lead and walk around with us on the homestead. Our time has paid off and now she will follow you around like a puppy. Bailey is two-years-old now and it's time for her to start breeding. She is our only full-time cow, so we decided to use Artificial Insemination (AI).

Injection Time!

Since we spent so much time with Bailey during her life, Ron and I hoped that it would be a simple thing to give her an injection. Our veterinarian Cody Ames, D.V.M. of Tonasket, gave us two syringes ready for Bailey.

It was up to us to give Bailey the injections which start her heat cycle; it is then that she can be artificially inseminated. The cow has to be in heat at the time AI is performed.

♥ Of course, nothing is ever simple here. Ron had to give Bailey the injection because women or men who have Diabetes are not allowed to get in close contact with the syringe at all.

♥ I was told not to handle the shot. I'm not pregnant, but if you are pregnant the veterinarian won't even allow you to handle the *box* that the syringes come in!

♥ It's a pretty powerful shot that gets the heifer cycling so she may be bred. At any rate, Ron doesn't have Diabetes so he could give Bailey the necessary shots.

I went out to help. First, I told Ron about giving shots since he hasn't had to give one before. These syringes are a size eighteen, they're thick and you have to push a bit harder to get it through the tough skin of a cow.

We closed the door to Bailey's stall so she couldn't run away from us. It was about twenty degrees outside and I was really cold. I wanted to be inside the house by the wood stove!

♥ I brought Bailey a pile of hay to keep her busy and Ron jabbed the needle in her neck where the muscle is.

♥ However, he didn't have time to pull back on the syringe to check if there was blood in the needle because Bailey came unglued.

Bailey trusts us, you see. We've never hurt her before. And I guess it must have hurt for that needle to be jammed into her neck. She did a couple jogs around the stall and the needle worked its way loose and fell to the floor.

Of course, I bent down and grabbed the needle so Bailey couldn't step on it and hurt herself. It was my natural instinct.

Ron yelled at me since I'm not supposed to touch it but hey it was instinct and who can stop that? Luckily none of the injection had seeped out so it didn't absorb into my skin or cause me any discomfort.

♥ We decided we needed to make a cow squeeze chute for Bailey to be trapped inside so we could give her the injection.

♥ Don't make the mistake of thinking your cow is happy and tame and will just let you inject her with a needle. Cows have feelings, too!

♥ You may not get a second chance to use the needle if the cow gets wise to your intentions. Use a squeeze chute of some sort and get it all done in 'one shot'. Ha-ha.

For other ideas and examples on how to build a squeeze chute for your cow see *Storey's Guide to Raising Beef* in our resources.

Ron grabbed our extra 10' gate panel and we used bailing twine to secure it to Bailey's outside pen stock panels. This way we could toss hay up at the front of the squeeze chute. The area inside the squeeze chute is two feet wide.

♥ After we coaxed Bailey inside the squeeze chute to eat the hay, we pushed the panel as close to her as possible, almost squeezing her inside so she couldn't back out if she tried.

♥ Ron decided to use Bailey's buttocks for the injection site this time. He jabbed the needle in, pulled the syringe back a bit to make sure there was no blood in it and then quickly gave the injection to her. Bailey barely noticed the whole event!

This is how we will give her the second shot in two weeks. It should go smoother the next time since we have the squeeze chute ready for her and we know what to expect.

Three days after the second injection (which will put Bailey into her heat cycle); Cody (our vet) will come up and perform the process of artificial insemination.

The reason for two shots, states Cody Ames, D.V.M., "In case the first shot doesn't start Bailey's heat cycle, the second one *rarely* misses putting her in heat."

Cody Ames, D.V.M., Tonasket, WA

♥ Cows have a long pregnancy, like humans. Nine months and seven days. Or, 280 days.

The entire reproduction process is incredible to me. I won't pretend to know much about it, since Bailey is our first heifer! But I can't wait until next September when we can see our first calf born!

Storey's Guide to Raising Beef gives a lot of in-depth data covering the reproductive system of the cow.

Cody Ames, D.V.M. Performs the AI Process

I was a bit amazed at the ease of artificial insemination. Bailey was coaxed into the corner chute with grain and hay. The stock panel would immobilize Bailey while Cody performed the AI process.

♥ First, on went the glove. And what a glove that is! The length extends well past the elbow. Then Cody got down to business.

♥ He had to check Bailey's uterus to see if any eggs had dropped. Cody said he could feel at least two eggs. I was glad. That told me things were going along just fine.

Bailey just kept eating her grain and hay like no one was shoving their arm up to their elbow into her privates!

Now Cody went to his truck to get the semen. Upon arriving he had stopped up to the house and put some warm water in a thermos for this part.

♥ The semen has to be a certain temperature when placed inside the cow.

♥ Cody took the semen and inserted the straw into the warm water. Then he did an amazing thing! With both hands balancing his tools, the semen straw and Bailey's tail end, he inserted the straw into Bailey's uterus and the semen from a Black Angus cow was dispersed into her uterus.

Now we just have to wait for nine months plus seven days! I kept a pregnancy diary for Bailey...

Bailey's so-called Pregnancy Diary

The First Trimester

There were no differences in our cow during the first three months. We fed Bailey as I mentioned in the beginning of this chapter.

The Second Trimester

This time was the same as the first three-months of pregnancy. We fed Bailey the same and watched to make sure her water was full. And we bought her a new salt lick.

6 Months Pregnant:

Bailey escaped her pasture! She took off with a big crowd of cows and left us for six days.

We had a friend building a pole barn here, Nick Baker, and he noticed Bailey pacing the fence line. She was interested in the cows grazing outside of our fence. The next day Nick let us know Bailey was gone!

It's Been 9 Months

We are waiting patiently for Bailey to pop that calf out. I want to be there to help her. I expect since this is her first calf she will be a bit scared when the time comes.

In the Last Month

We've seen the most changes. During the first six months Bailey was fed as usual. She has 10 acres of grass and unlimited water with a salt lick.

♥ During the 9th month we've seen some changes in her development. Bailey's belly is enormous. I'm not sure how the calf will escape from its warm haven.

♥ When Bailey walks toward you her belly is approximately 6 inches outward on either side. (I guess this could be from our feeding her so well).

188

♥ Sometimes we can see the baby move now too! If you watch her stomach down towards the belly area you can see little kicks and movements that could be the calf.

Ron and I have been tossing Bailey grass hay sections for a couple of weeks now. We brought her up into a smaller part of the field. She doesn't have the entire field access. We need to keep a closer eye on her now that she is nearly due to calf.

♥ I don't want her to have problems and then not be able to give her the help she needs.

♥ Bailey has a small section of field for grazing with the 12' x 12' barn stall for protection from the weather, and an outside paddock to hopefully do most of her duties in.

♥ Of course we tossed out a lot of fresh straw for her to calf in.

From behind Bailey you can see her vulva swelling and start to droop a bit. Bailey's body is starting to prepare for calving. I'm so excited I can barely do anything else but sit and talk with her.

♥ Bailey is only two-years-old and has a small milk bag. She will have enough milk to feed her calf but right now you can't see how that is.

♥ Since Bailey is due to calf in three days, I check on her twice a day.

Passing her Due Date

Bailey is two weeks late for calving! She doesn't seem to be making any more bodily changes. And the changes we have witnessed thus far *could* have been normal for her growing body.

Ron says Bailey didn't conceive with the artificial insemination. I didn't want to believe him right away but it's been over nine months now and Bailey still has no calf.

Possible Good News

Remember when I told you Bailey escaped our fence at six months in her pregnancy? Well, she might have been in heat then and smelled a bull. We still might have a calf. We just have to wait another five months to find out.

While I'm waiting to find out if I will ever get a calf from my cow, I will tell you about a few other cows which have already had their calves. Then I'll update you on Bailey's process.

Sheri's Heifer Calves

I've mentioned my friends Bill and Sheri Vejrostek. On their farm they started out with one heifer to breed; now they have about thirty at certain times of the year.

They live on 'Wes Ranch' which is Sheri's grandfather's legacy. Bill says he will see to it that the legacy continues. If you want to buy cattle, either heifer's or calves, hogs of all sizes, quarter horses, dogs, ducks, chickens or just a dozen eggs, this is the place to go!

Our oldest daughter has raised a puppy from Sheri's farm and we brought her over to visit. Jynx is five months old now. She is absolutely beautiful. And of course so is our daughter, Shana. Shana has been trying to teach Jynx tricks and obedience.

Shana with Jynx

It was a sunny spring day when we came to show off Jynx. Sheri had mama cows about to calve and I wanted to see what to expect if Bailey had a calf.

♥ Sheri showed me how the vulva on a heifer about to calve will enlarge to five times its normal size.

♥ It was extremely swollen and drawn downward in the same line of the tail position, which at this point was still straight down.

191

♥ Soon the tail would bend over to one side to allow the baby to emerge.

This cow was in an outside pen with the other pregnant heifers. Mothers will leave the herd to calf in private.

After she cleaned and fed her baby, the cow came back outside to join the others! It is the mother's instinct to use the protection of the shelter to have her baby.

Cows have good mothering instincts. I think they get a bad rap most of the time for their wit. Heifers are far from stupid! Another one of Sheri and Bill's heifers made her way into my book with her beautiful baby girl. Sheri had been watching and waiting with her camera for this baby to emerge.

♥ The mother seemed to be holding the birth back, though, waiting for Sheri to leave.

♥ First you see the front hooves emerge from the long line of a hole the mother's vulva has become. It's really neat.

♥ The calf's hooves are almost a white color at first.

♥ The mother's tail is off to one side to allow the calf plenty of room to emerge into our world.

♥ Baby is covered in a wet goop that only a mother could love (and lick away).

♥ If healthy and well cared for most of the time cows need no intervention from us farmers.

Heifer licks her new born calf clean

Here is a picture of the new mother licking her baby clean. It's hard to tell with a black and white picture but there was a thick wet film and blood covering the baby. This is a female calf and she was drenched from her own birthing bath.

Soon she will be dry and on her feet looking for a drink of milk from mama. It's important that she get the colostrum her mother provides for her with the first few days of milk. Spring sure is a wonderful time. I am more eager for Bailey to have her calf now!

♥ Don't forget when your cows are out in pasture that they need water. Cows should have water in front of them at all times, especially in warmer weather. One nursing cow can drink 35 gallons of water a day!

Bailey's Progress

Bailey just isn't going to have a calf this year. I'm even more disappointed than you must be after reading this far. But the proof is here. We just witnessed her coming into heat and having her time of the month.

In case you have never seen this happen, the cow acts normal except she oozes blood from her vulva. I actually saw blood oozing out of her vulva when Bailey was lying down and went to stand up. She could use a diaper for sure!

The blood is bright red, thick, almost like a mucous and it gets all over her tail which then spreads the blood across her backside.

Our Butcher Steer

Ron bought three Holstein steers from a dairy farmer on his way home from Moses Lake one week last September.

I believe once you have a herd started and the heifers are raising the calves for you, then the outcome could be profitable. But not for us. We paid $130 per head for the three steers and we had to raise them ourselves.

They were all three castrated males, so we couldn't even add to our herd with them! It wouldn't seem so bad to me if they had bonded to us but since there were three of them they kept in a corner most of the time bonding to one another. They would come to the fence to smell our hands, but still wouldn't give us kisses or bellow out a loud 'hello' when we come to the barn, like Bailey does.

After two and a half ton of alfalfa hay, one ton of grain and seven months of feeding and watering the steers we are going to make a grand profit of $65.00 per head. We are butchering out the third calf for our freezer, though. So that is one good point.

The kids and I enjoyed raising the steer. They are easy to take care of. But we like the animals that we raise from birth and become part of our family. I do want to buy a few more meat cows but I don't want the castrated males because I don't see any reason why we can't buy all females just in case we get a *really* nice one we want to keep for a breeder! (Don't tell Ron that last part, though!)

Dogs Attack Cows

This one is a little closer to our hearts. For seven months we raised and loved three Holstein steers. It was time to sell and we had a buyer. Nick Baker of Aeneas Valley, Washington.

With all the paperwork done we hauled the two steers down to Nick's field. The only problem was our third steer (who was much smaller) joined the other two. We couldn't separate them while loading, it was too difficult.

Nick said, "No problem. You can keep them in my field."

"Are you sure?" Ron asked. "We're going to have him butchered this fall."

"Sure." Nick told us. "There's lots of grass for them to eat. It's no problem. When the slaughter truck comes, I'll just tell them he's yours."

So we left our little guy in Nick Baker's field. Nick had never had a problem with cows in his field during the fifteen years he had been raising steers.

Two days later I got an email from Nick stating that two dogs had attacked the cows and our little guy was pretty mangled. Sure hurt to hear that. Seven months of bottle feeding, graining, then finally our steer is on full hay and he's brutally attacked by dogs!

I drove down to the field immediately to see how bad the damage was to the cow. Nick said he hadn't seen him eat hay with the other two since the attack. The other two steer had a few scrapes and bruises and the dogs had run them through a barbed wire fence before going back to turn our little guy into hamburger.

When Nick got the call from his neighbor he grabbed his gun and hurried out to his pasture. What a sight it was when he got there. There were two dogs on our little guys' face. Nick shot one of the dogs right away. The other dog ran off. (Nick shot him later).

Nick told me the next day that after he shot the first dog he almost shot the cow, too. The poor little guy didn't even jerk or jump when the gun shot went off next to his head. Nick sat down with him and the little cow laid his head in Nick's lap!

Nick sure didn't want to call me but he did and I drove down quickly. Within twenty feet of our cow I could see all the blood dripping down the front of his white and black legs. I choked back a sob when I could hear him breathing from the same distance. Apparently the dogs couldn't get at the two bigger cows very well so they dug into my little guy who was only four hundred pounds or so.

The steer had a broken back leg with deep lacerations from bites. Broken ribs which made him wheeze and heave as he struggled to breathe.

Like I'd heard done in the past, the dogs took my cows' cute little black nose and chewed it right off his face. Now instead of that nice black wet nose with two holes to breathe through, all I could see were two white bones where the holes should be covered with skin. It was heart wrenching.

I told Nick I would have the slaughter truck come out ASAP and put the steer down. I didn't want him to suffer. Poor little guy. And poor us! We bottle fed that cow, we raised him on grain and alfalfa hay and then we were planning to raise him another seven months before butchering. This would have been our family's first fresh beef. It was the biggest bummer.

Yes, we did have the steer butchered into 168 pounds of hamburger. But there wasn't any of the nice marbling in the meat since the steer was so young. The meat is rather blah.

Why did the pig ask the farmer for some sun screen? She was bacon!

16
Funny Farmyard stories

Bailey: Our cow

Everyone gets a kick out of this story! I had never been to the Okanogan County Livestock Auction. My friends -Wendy Mahlendorf, and her daughter – Melissa Maple, (and her daughter Miranda) took me to the auction to waste a few hours one Thursday afternoon. Being October, it was cold. I had Kyra along, and she played with Miranda on the seat next to me.

The only thing Ron told me before I left was; "*Don't buy anything.*"

First, we devoured our hamburgers. Wendy had bought some m&m's for the kids. Melissa didn't want Miranda to have candy, but Wendy said it was a learning experience; she asked the kids to tell her what color the candy was and if they knew the color, they got the candy. So, Wendy was teaching the kids their colors, right? Who could say no to that? Not myself and not Melissa after she thought about it.

Now the sale started. People, mostly old farmer men, were flocking in, taking seats. One of them was the owner of the little female day-old calf I wanted. I tried to figure out which one, but none of them looked like her. You know the old saying; owners start to look like their animals? Not so with cows, I guess.

You must think I was bored at the sale with all this thinking going on in my head, huh? *Not at all!* I was having a blast! Too much fun as a matter of fact. The auctioneer was calling out his bids. I didn't know *what* he was saying!

When the big door slid open and a beautiful cream colored day old calf wandered out, I sucked in my breath. My mouth was open wide when I sucked in my breath, though, and it was loud! Wendy jabbed me in the ribs with her elbow. Melissa leaned forward to see around her mom and told me to 'shh.'

Folks down in front of us turned their heads. Cowboy hats tilted and turned to look our way. I couldn't help myself! The calf was the prettiest color. To look in her eyes you would think she was a baby deer; she was so pretty. At twenty feet away I knew I had to have her. And I couldn't be quiet about it either.

The auction is like the library. You are supposed to be quiet and listen.

"That's my cow!" I said in awe. It came out louder than the gasp I let out just before. I didn't plan to buy anything. I was told not to, remember? My arm went up pretty fast though, without my control, and she was ours!!

Ron teaches Bailey to lead

It took me four hours after arriving home to get nerve enough to tell Ron what I'd done. The last thing he had told me before I left was, *"don't buy anything."* During the hours I was trying to get my nerve up to tell him, I kept hinting how nice it would be to have a baby calf.

Finally, he caught on and sarcastically asked me, "What did you buy, Jeanie?" He knows me too well!

We had to load up Big Red with straw to go pick up the calf at Melissa's house in Riverside. Her husband was just as surprised as mine when he got home; she had bought two calves! I hoped *that* knowledge would make Ron a bit less upset with me since I only bought one calf.

What upset him most was that he was planning on relaxing and now we had to go driving around half the night. Melissa lives an hour away from us.

Ron and I put our calf in the back of Big Red. I was worried that the calf would catch a chill but she was fine.

We did have one problem. It was pretty scary, too! This kind of stuff always happens to me. Ron is used to things being normal. So, he wasn't ready for it when the head lights suddenly went out while we were driving home on highway 97. It scared me, but I reacted right away since this stuff happens all the time to me.

I yelled, "Step on the high beams!"

Ron did. The high beam lights worked. Ron tried the regular lights and they worked now, too. Just another reason for him to yell at me because I bought the calf. You see, we wouldn't have been driving home in the dark on highway 97 when the headlights went out if we didn't have to pick up a calf I bought when I was told not to, right?

It's a tangled web we weave, but it always works out for the best, though. You know, everything happens for a reason. (I still have no idea why the headlights went out, though.) As soon as Ron saw our calf, he fell in love with her just like the rest of us did. We named her Bailey because she is the color of Bailey's and cream.

At two weeks old Bailey was being halter broke. At first she would have nothing to do with it. Pulling back hard, she could almost yank you to the ground! Eventually she caught on and gave in to the lead... Now she is over 900 pounds and we can lead her around!

Bailey didn't always like to walk with us!

She is two-years-old now. We have halters for her and we can lead her around the place like a dog on a leash. Bailey is very friendly. The girls are always trying to ride her. Bailey will just walk away like it bores her.

Old Timers used rabbits/chickens to Clean

Did you know that the old timers used rabbits or chickens to clean their fireplaces? They didn't have those fancy wood stove brushes back then.

- ♥ What they did is attach a chicken or rabbit to a rope and lower it down into the chimney. The animal would then scratch, kick and jump.

- ♥ This constant movement from the animal would loosen the creosote from the chimney walls.

Pretty inventive, huh? People would get thrown in jail these days for such animal cruelty!

Don't use Diesel when you need Regular

Ron knows me well and there's a bit of a joke about how I am the 'queen of an empty gas tank'. If you need to know how empty you can get any of our vehicles before it runs out of gas, I can tell you what to watch for. I am good at this. Sometimes I coast the last few miles to the gas station on *fumes alone!* Luckily for me it's all downhill the last six miles or so into town.

I've been warned and preached to often enough that if I look at my gas tank before I go to town and there obviously isn't enough fuel to get me all the way there, I will try to get there anyway rather than listen to Ron pester me about it. I don't like the 'I told you so' look he gives me either! It's just too much.

Amorette and I had to go to town one day and as we were leaving I noticed the gas tank read empty. It was already on the red line and the fuel light was on!

I weighed my chances of making it all the way to town or asking Ron to dump in a few gallons from our gas cans we keep for the tractor.

Amorette agreed with me that we could park next to the shed and put in a few gallons from the gas cans and Ron would never have to know! So, that's what we did.

Have you ever heard that saying, 'What comes around goes around?' Well, the saying came to life for me here.

I backed up to the shed and nonchalantly grabbed the gas can and started pouring its contents into my Jeep. What I found out when the Jeep died before it drove a quarter of a mile out our gate was that the gas was not gas at all. It was diesel fuel!

I did a big giant no-no. And I'll tell you this much, I will never hear the end of this one!

When I told Ron he just about blew his top! Cussing under his breath, he turned his back on me and walked away. I couldn't believe it.

Was it my fault that he put diesel in the gas can? No. And when I yelled this knowledge to him, he let me have it some more!

"You did that, Jeanie! Remember when you went to the Little Store to fill the gas cans and Dink asked you if you were sure you wanted diesel?"

Oh yeah. I did do that. Am I the only one around here who felt like a miserable screw up at this point? Yes. Ron wasn't going to let up either.

"Now you probably ruined the Jeep. The whole engine will be toast now! I'm not a mechanic, what can I do now?"

Ron was pacing back and forth in the yard. I could tell he was really ticked off. Under his breath he was cussing, but out loud he was trying to smile and seem courteous.

That's when you know Ron is angry -when he tries to smile while he gives you a what-for!

I sat inside the Jeep with my tail between my legs and would have started crying if Amorette wasn't still sitting there. She was grinning from ear to ear!

"I guess Ron knows we're out of gas, huh?" She said.

"Yeah, Amorette. I guess he does. What's so funny?"

"I just think it's funny that you pick the *one* gas can with diesel fuel in it. All the other cans are filled with gas. But *you* picked that one! That's what I think is funny!"

"Yeah," I said, *"that's just my rotten luck!"*

I wondered why we couldn't just drain the tank, siphon the diesel fuel out and start over.

After I thought Ron had had enough time to cool off, I went over to the barn to ask him about it.

He didn't think we could do that, he thought I had ruined the Jeep. The fuel jets would be clogged and damaged by now. I argued that it had *just happened*, how much damage could there really be? He laughed and we walked over to the Jeep.

"Humor me, Ron." I said. "Let's empty the gas out and flush the system with fresh gas."

"You mean *real gas*, don't you?"

I just gave him one of my evil-eye looks and he smiled and turned his back. He knew he won. He knew I knew he won. That's what I hated the most right at that moment.

Sheri's Farmyard Runts

I've mentioned before that Sheri Vejrostek's animals rule her farm, and that is true. We visit Sheri and Bill's place as often as we can. Every time we do we see things that amaze us. Here are a couple of other facts I didn't mention!

One time (during the summer) we came by to look at the sow we're buying. I wanted to make sure she was pretty looking and good size and what we want our sows to be. I knew she would be tame since she would be coming from Sheri's.

Well, Kyra had to go potty, she was potty training, and so we needed to use the bathroom. Sheri took us up to the house. She warned us as she usually does about the order of her house.

"Don't look at my house. It's a mess. I've got babies all over the place!"

I said not to worry about it, we didn't come to see her house, we came to see her, and besides our place was a mess, too. No big deal. We're used to it.

We didn't give her mess a second thought. What we were thinking about for days, weeks and months later were the piglets that seemed to know where they were going as they wandered in the house after us and took a seat in front of the TV!

"What are they doing?" I asked Sheri, "Are they going to watch Babe or something?"

At this time Sheri had three runts that had disowned their mama pig and adopted Sheri as their mom! She would fill their pan out front with goat milk. They ate heartily. They did play in the yard with the other piglets, but when it came time to eat, they made their way to the house, and actually whined at the door to be let in!

The other thing we can't forget was the mama dog. She had nine babies on a blanket in Sheri and Bill's bedroom. It was awesome. The babies were just a week old or so, and my little girl couldn't get enough of them! She squealed in delight.

"Mama!" She screamed to me, "I need one of these!"

"Why, Kyra?" I asked her.

"Because! It's my best friend!"

Kyra didn't get one of those baby pups. My neighbor, Wendy Steever, is the owner of Howling Ridge Rescue for Dogs and she would have killed me. She rescues unwanted and dropped-off dogs for a living. She always has a dog to sell. If we were to pick up a pup from someone else, she'd probably have me bush whacked!

Just when I thought it couldn't get any more exciting, two of the runts came over to the mama dog and started nursing. And the mama dog laid there and let them! It is always an adventure over at Sheri's.

Sheri...is she crazy or just nuts?

Sheri could very well be crazy! Who knows? When Sheri visits her daughter & husband on the coast her son-in-law sees her and says "Oh my God, your mother's here!"

They think she's nuts. The last time she showed up she brought a sick baby chicken that she was nursing along at the time!

When she visits her sister it's always a nutty time. Sheri arrived at her Prim & Proper sisters house (with beige colored carpets and white curtains) carrying a runt piglet. You can imagine how the trip went. The piglet was not potty trained! Her sister thinks she has completely lost her sanity.

Bill and Sheri also have a pot bellied pig they named Samantha. Samantha was bought as a two-week-old and is now four-years-old. The funny part about this story is that Samantha is an *inside* pig. She eats what Bill and Sheri eat, and she sleeps with a pillow and a blanket on their bedroom floor!

Stolen Firewood

When a friend of Nick Baker had missing firewood over a three year period of time he got angry.

"Why do people steal firewood?" he asked.

Who knows? Maybe they are just too lazy to chop and split their own wood or maybe they can't afford to pay for their wood. Either way, this friend of Nick's wanted to know who was stealing his wood.

He set a trap to catch the perpetrator. Carving a log open so that he had a hollowed-out area inside, he inserted an M80. Whoever was stealing his wood would have a big surprise!

About three days later he heard a huge explosion from his neighbors' house. The M80 had exploded and blew up the fireplace! He was shocked to find out the thieves were *right next door!* But now he knew who the perpetrator was.

Soap & Water!

There is always one of those animal stories you hear that you just can't forget. This was the story for me. We all get a huge laugh when we remember Soap & Water. It is all based on a true story.

There's this guy who lived alone in the woods, residing in his cabin. He kept to himself a lot. His sister lived in the city and once a year she would come to visit.

They had a nice visit. His sister was curious how he lived like he did, so far out in the country. And he was amazed at how she could live in the city. He made them a nice dinner and they visited for hours.

When they were done with dinner his sister tried to collect the dishes to wash. He waved her off with his arm, "No, sit down. Soap and water can do that."

She didn't know what he meant. She tried to collect the dishes again, saying that she didn't mind.

"You cooked dinner, so I'll wash up the dishes."

But her brother wouldn't hear of it. He told her to sit down; he would take care of it. He got up from the table after she sat back down, saying, and "This is how we wash the dishes out here." He walked over, opened the door and called out, "Soap and water!"

Bolting through the door, jumping up on the chairs the brother and sister had just sat in to eat dinner, and licking the food remains from the plates, were two mangy looking dogs!

Needless to day, his sister was mortified that he used his dogs to clean dishes, and literally sick to her stomach to think that she had eaten from those plates! She left to go home the same night.

Keeping a log-sheet at the barn can make it easy to jot down important records such as when bred and to whom, due date, when birthed and number of babies, etc.

17
Briefly: Keeping Records

Keeping good records is the key to the start of good business sense! We keep *many* different types of records on our homestead.

♥ First, health is of the most importance to us. Our healthy animals are happy producers.

♥ We routinely give vaccines and worm our livestock on an annual schedule which works great for our area.

You should talk to your local veterinarian to see what vaccine and worming schedule would work best for your area. Prevention is much easier and cheaper than treatment.

Newsletter for Customers

I created a quarterly newsletter for my customers. The newsletters have a lot of information concerning our homestead, pigs, farm facts, recipes and just lots of good stuff. We even have contests through email and the folks win prizes!

The newsletter keeps my customers informed and in touch with us. It's great advertisement! You can subscribe to our Newsletter on the last page of this book.

♥ I have now added a new mailing list to my web page which sends you an FREE email newsletter every once in awhile. This is free to anyone who goes to my web site and adds their email address to the mailing list!

♥ Once a month my free mailing list newsletter randomly selects an address from the list of email addresses, and the winner receives a FREE book!

Subscribe to other magazines or newsletters that will help you keep in touch with the folks in your business. It helps to be able to ask questions from someone who has been through it before.

Introduction to the Basic's

Of course this is all just my opinion and the way I do things on our farm. I'll list what I've learned and you can send me information to clear up any of the blind spots, okay? I learn new things every day.

I do feel records are of the highest importance on a farm of *any size*. Who can remember from one year to the next which pig had fifteen babies or which goat had triplets regularly? Not me. I write it all down.

♥ Even if you have a great memory, chances are your memory won't be as bright in five years.

♥ And after keeping records for five years you can start to make better business decisions for your farm.

I will list the highly recommended records we keep. I store the data in a two inch binder for easy filing. There are charts I have copied from other books and then reconstructed each to suit my own purposes. You can use mine as a basis and then change them to fill your own needs.

For a complete report with blank copies of records and sheets you can use right away, see our document #275 *Small Farm Records* in the resources section of this book.

♥ Happy, healthy animals don't get sick every time the wind draws a virus. This is why I vaccinate and worm on a regular basis.

♥ My animals rarely get sick. When they do it's usually due to droppings from a strangers boot or by an animal coming to our homestead for the first time.

Just-A-Little Ranch			
Ron & Jeanie Whiting	**Name:** Bailey		
P.O. Box 1516	**Sire:**	**Dam:**	
Tonasket, WA 98855	**Foaled:** 10/99		
Worming/Shoes	**Description:**		
6/1 worm-Panacur #	**Owner:**		
8/20 worm-Panacur			
	DATE	VACINATION	ANTIBIOTICS
9/00 Ear tag -	9/00		
8/10/00 →	8/10/00	Bangs Vac.	
		Clostridium	
		C+D Toxoid	
8/12/01 →		Bangs Vac	
		Clostridium	
		C+D Toxoid	
9/ /			

Health Sheet #1

Now, depending on what your farm is trying to accomplish, you should keep records for health checks, worming, vaccines, trimming hooves or shoes and list any accidents that occur; per separate animal.

My health sheet is three pages long. I keep them in a two inch binder for easy filing and retrieving information.

Health Sheet Storage

♥ I keep a two inch binder with all of the animals' health and breeding records filed inside.

♥ Each type of animal (pig, cow, and goat) has its own section in the binder.

♥ Then I sort the animals by name (Rosie, Bailey, etc.) And it's a good idea to keep a picture of each animal in case it is lost or stolen.

In the health binder I keep records of each time we worm, vaccinate or breed. If there are any accidents or injuries this is where I list them.

♥ I'd like to know in five years which animals cost us more to vaccinate or to keep healthy.

Vaccination History

Date	Disease	Drug Used	Dosage	Date	Disease	Drug Used	Dosage

Parasite Control

Date	Method & Drug Used	Date	Method & Drug Used	Date	Method & Drug Used

Injury or Illness

Date	Description or Nature of Illness	Treatment

Health Sheet #2

Health Record

| Name: Katie | | Birthday: 1997 | Sex: Female | ID # 1 |

Color & Markings: white, horns removed

Owner:

Katie — Alpine Sire / Saanen Dam

Breeding Record

Bred To	Bred	Due	Delivered	Sex	Name	Color
Saanen	#9/00	3/23	2/24	3		White
Saanen	10/01	3/24	3/25	4		White

Health Sheet #3

Breeding Stock Records

I keep a separate sheet (per breeding animal) to track when the animal is bred and to whom, when due, when birthed, how many alive and dead, and how many males or females. This breeding sheet is filed after the health sheets in the binder, per individual breeding animal.

Just-A-Little Ranch
Ron & Jeanie Whiting
PO Box 1516
Tonasket, WA 98855

Breeding Record - Rosie						
Date Bred	Name of Boar	Date Due to Farrow	Date Farrowed	# of Live */Born	# of Gilts	# of Boars
Nov-6-1999	Gordy	2/28/00	2/27/00	8*9	4	4
4/21-22/2000	Abraham	8/14-8/15	8/15/00	12	3	9
10/26-27/2000	Abraham/Gordy	2/18-2/19	2/18/01	14*15	5	9
10/2001 ?	Abraham	Feb 2002				

For our sows we use a simple breeding record that keeps all of the breeding information together in one place.

♥ You can convert this form to use it for your cows, horses, or whatever animals you breed.

Keeping a Milk Record

The breeding sows (or other breeding animals) won't need any more than the records listed above, except maybe a picture of each one in case he/she gets lost or stolen. But I have milk goats and need to record the weight of milk received daily from each goat.

♥ I believe you need to keep records of the weight in milk you receive from each goat on a daily basis.

How else will you know who is slacking off? I keep a simple milk record sheet up on an easily assessable board in the milking parlor.

♥ After weighing the milk I write in the weight of milk received from each milker. We have three milkers right now.

♥ With this record of daily received milk I can keep good track of what is going on in my barn!

♥ One gallon of milk equals eight pounds.

I add up the daily weights at the bottom of the monthly sheet. This way I have a daily and a monthly record. And it doesn't take more than a few minutes a day to do this.

Monthly Milking Record

Month:	April	Katie	Eve	Shyann	Comments
1		2.5 / 1.5	2.3 / 1.1	1.8 / 1.3	44 degrees/bought 50# bag grain
2		2.4/1.6	2.3/1.2	1.8/1.2	44°
3		2.6/1.7	2.2/1.2	1.7/1.3	42° Sunny
4		2.5/1.5	2.3/1.1	1.8/1.3	44° trimmed hooves
5		2.4/1.6	2.3/1.2	1.8/1.1	42°
6					
7					
8					
9					
10					
11					
12					
13					
14					
15					

Milking Record Sheet

Now at the end of the year, I should be able to add up each month's milk received from each milker and get a yearly total. This will be exciting for me to see.

You can see how over the years I will be able to assess which milker is still doing her job and which are slacking. Records are important if you want your homestead to produce for you and prosper!

I used the basis of a nice milk record sheet from a favorite book of ours - *Storey's Guide to Raising Dairy Goats* and converted it to work for my barn. I also ordered a pad of milk record sheets from a *Jeffers Livestock catalog*. But I like to use the milk record I converted better.

You can order a catalog by calling 1-800-Jeffers. Their catalog has many items and things we need.

Family Tree Chart

Just for fun, I like to create a family tree for the animals my customers purchase. It's just a bunch of fun for me. You can change this chart to fit any type of animal you breed, also. I feel it shows my customers that I care about the animals I produce and sell to them.

The chart simply lists the name of the animal, birth date, breed, sex, color, and weight. Then I use as much data as I have available on the animal to list its parents, grand-parents and great-grandparents, along with any given breeds.

Breeder: _____

Family Tree

			Great Grandparent: Sire: Breed:
		Grandparent: Sire: Breed:	
	Sire: Color: Wt.: Breed:		Great Grandparent: Dam: Breed:
			Great Grandparent: Sire: Breed:
		Grandparent: Dam: Breed:	
Breed: Born: Sex: Name: Color: Wt:			Great Grandparent: Dam: Breed:
			Great Grandparent: Sire: Breed:
		Grandparent: Sire: Breed:	
	Dam: Color: Wt. Breed:		Great Grandparent: Dam: Breed:
			Great Grandparent: Sire: Breed:
		Grandparent: Dam: Breed:	
			Great Grandparent: Dam: Breed:

Date: _____
Sold to: _____
Phone: _____

Providing a Family Tree Chart for your customers will let them know you care about your animals enough to keep track, and it makes selling more fun!

Keeping Track of it all

The last thing I want to mention is a computer program we use to track all of the grain, hay, equipment, livestock and other supplies we buy for the homestead.

Quick Books Pro has worked for us for years now. You can track many separate businesses with this program. The computer does much of the work, especially the math.

It makes it nice for folks like me who can't find the time to work numbers every week. I just type in the information and the checks we write and the computer keeps all of the necessary information our CPA needs.

At the end of the year we print out a detailed copy of the profit and loss report and bring it to our CPA for tax time.

Did you know that pigs can eat a human being flesh, bones, clothing and all?

18
Scary Farmyard Stories

Beware of the Mean Sow

This is hard for me to add to my book because I love pigs so much and I don't want them to have a bad name. You will hear horror stories concerning mean sows, or boars. I have to add this section for parents to be perceptive of pigs of any size until you know that they are gentle, and then, always be cautious.

Ron, me and the kids never know if Rosie is having a good day or a bad one so we always approach her through the fence first with a nice scratch to her head. She's *never* been foul with us yet, but if she was going to be we hope she would give us a sign at this point.

225

We don't let Kyra into the pen with Rosie alone, Kyra is only four-years-old; Rosie is 800 pounds or more. It's only common sense, right? We encourage all parents to be cautious when children are around pigs, and make sure the kids are educated about what *can* and *does* happen.

For this book, I asked neighbors, friends and anyone who knew a scary story about animals to tell me about it. I just wanted to put a chapter in here on the dangers of what could happen. People don't usually tell you the bad things that happen. I haven't read a book yet that had these true stories of mishaps and losses in it so I thought it would be good to add.

A Pet all her Life

Terry Dean had a sow she had raised since the pig was four-days-old. She fed and loved the sow for years. Well one day the sow surprised her and she will never forget the experience.

The sow was in a hog pen next to Terry's goats. She had been next to these goats all of her life on the farm with Terry. Of course the sow had never been mean before and that was good because she weighed 570 pounds.

Well, one day Terry went out to check on the herd and found a goat's head was stuck through the fence. The goat had stuck its head through the fence, into the sow's pen, and the sow had eaten the poor goats head right off its face!

And the part that Terry will never forget is that the goat was still alive! All that was left of goat were the bones on its face. The goats' body was still intact on other side of fence!

Sows Meeting with Litters can be Fatal

Two nursing sows came together at one farm and there were fatal results. The two sows had been enjoying their time with new babies when they came face to face with each other.

As the sows fought they went around and around in a circle, and seven babies were stepped on and crushed to death.

Dogs Attack Goats

There doesn't seem to be much use for dogs that attack livestock. Maybe that is why some people have them killed.

I wish it were possible to tell everyone how important it is to keep *your own dogs* on *your own property* even when you live out in the middle of nowhere. Some folks let their dogs roam the area. And the owners never take responsibility for what their animals do.

This is what happened. And it was sad! Four dogs were roaming the area. Their owners had left for the coast (over six hours drive from here) and left the dogs loose. As roaming dogs will do, these four had been chasing folks' livestock around this area for months. People had asked the owners to keep their dogs tied but the owners would not take their responsibility seriously. They leave it up to us farmers to take care of *their* dogs!

My neighbors Galen and Shannon had asked the owner to keep his dog's home more than once. This is the response they received:

"It's the rule of the land, Galen. If my dogs bother your animals you just shoot them."

That is the biggest cop-out that I have ever heard! Who tells a neighbor its okay to shoot their *own dogs?* If they care that much for the animals why even own them? It's infuriating.

Bright and early one morning a few months after the above conversation, Galen and Shannon were awakened by their animals screaming for help! Shannon jumped out of bed and ran to see what was going on. She opened the door to see the four dogs chasing their goats across the field. The goats were heading home to their barn for protection and the dogs were right behind them.

Galen grabbed his gun. By the time the two of them got to the barn the dogs had cornered the goats inside the barn. Damn dogs killed three full grown goats and two babies. And as if that wasn't enough they wounded two more babies! Galen was able to shoot two of the dogs but two of the four got away. Later he would hunt them down and end more dogs' lives.

Before the dogs had gotten the gall to chase and kill the goats they were busy with other livestock. After cleaning up the mess from the dead dogs, Galen found two dead roosters!

These dogs were not hungry, they just wanted to taste blood. Eventually the last dog was taken care of. The owners never took any steps to paying Galen or Shannon back for their lost livestock, or for the cost of the tooth Galen chipped when he slipped and fell while shooting one dog. It cost them thousands of dollars and they had to endure the loss because these neighbors are not only inconsiderate but they didn't have the means to make things right. Isn't that typical?

Goats who Couldn't Deliver their Kids

A fellow neighbor shared this story with me. She had bought goats from someone in the Valley and raised them up to breeding age. Finally they were bred to a buck. Goats have a gestation time of five months minus one day.

When the time came neither of her goats were able to deliver. I still don't know what happened or why the goats could not deliver their babies. But the lady and her husband took matters into their own hands.

Her husband went out and shot each doe in the head. Husband and wife proceeded to cut open the does' bellies and rescue the babies (each doe had one kid inside). The babies lived.

Protective Sow on Castration Day

Be wary of a sow which has been a pet all her life. It's true that pigs are friendly and wonderful to have around. But this story is close to our hearts and involves my beautiful Rosie. You have to have a great deal of respect for the sow when she has a litter.

First let me ask you a question. If you were in the Mall shopping with your small child and you looked around after hearing the child yell to you and saw a man running off with your child, what would you do? My guess is that you would take off after that man and get your screaming child back. Am I right?

Well, the day came for us to castrate Rosie's boys. Bill and Sheri came up to do the job. Actually Bill was showing Shana the process again as a favor. Rosie was let out to roam the homestead and hopefully get out of hearing range while we castrated her boys.

Having all the supplies at hand, Bill and Shana went to work. Shana caught a piglet and held it in position for Bill. The piglet was screeching endlessly and Rosie was right outside the pen pacing back and forth.

It doesn't take long to castrate, probably four minutes per piglet, and Rosie would calm down and go about her own business as soon as the piglet stopped making all the fuss.

Bill was showing Shana the ins and outs of castrating the second piglet when Rosie decided it was all too much for her. She wanted in to get her babies away from Bill. Rosie started nosing the pen from outside. I was standing there telling her it was okay, it would be over soon. These weren't the comforting words she wanted to hear and I was a bit nervous with her pacing like that. She is over 800 pounds, you know.

Well, it all came to a halt when Shana tried to catch up baby boy number three to castrate. She caught him all right but all the other babies went right through the fence running off to be with Rosie! I couldn't believe it. The opening was only about six inches wide and there was an electric fence there to zap them as they went through. But they all went to find mom.

We still had two more males to castrate. Of course, now that she had her babies back, Rosie jogged off in the other direction!

Sheri suggested that we let mom (Rosie) calm down for awhile and then try again. I wondered how we would get all those piglets back into the pen. But we did.

After awhile Rosie went in her pen to eat the grain we had dumped to lure her back inside and the babies followed her in. Bill decided it was time to try again. I asked Shana to go inside the pig house and bring out one of the two males we still needed to castrate.

When Rosie saw Shana come inside her pen she just freaked out. She paced back and forth and wouldn't let Shana inside the house. I tried to keep Rosie busy petting her and giving her treats while Shana tried to catch a male from inside.

Bill was also trying to keep Rosie busy, but he was upsetting her. Bill was outside her pen reaching down to pet Rosie, but she didn't like him because he had made her babies cry and she was trying to nip at him! I was shocked at this behavior.

Shana finally got a baby and she and Bill went inside the chicken coop this time so the noise wouldn't bother Rosie so much. We returned the baby as soon as the job was over but Rosie didn't calm down this time.

She didn't like Bill one single bit. Bill wasn't helping any by trying to get Rosie to bite him, either. He was reaching into her pen and she was jumping two inches or more off the ground to try and bite him. No more nipping, which was upsetting enough for me, now she was trying to take a bite. And Bill just kept baiting her to bite him.

I was pretty upset by now. I'd asked Amorette to take Kyra up to the house because I didn't know what Rosie would do anymore. She wasn't acting normal. She was being aggressive and panicking because of Bill's actions. Shana was still trying to catch the last piglet.

Rosie didn't want anyone in her house anymore. She chased Shana out and that scared me and Shana.

Bill said to try holding a piece of wood over the doorway so Rosie couldn't get in to chase us out. Shana went in the house from the other side of the pen.

Remember, we have one house divided in two sides with two by four boards to keep the pigs on either side. Rosie was on one side, and I was in the other pen next to her, just over a thirty two inch hog panel fence next to her doorway. Shana climbed over the two by fours thinking Rosie couldn't see what she was doing since I was holding the board over Rosie's doorway. The babies started squealing and Rosie came running towards me. She wanted in her house.

I held the board there but Rosie didn't stop. Rosie looked up at me and came *right over that thirty two inch fence!* Yes! She didn't even leap. Rosie just lifted her feet and came over that fence like it was a two inch barrier. I was startled, dropped the board and jumped over the fence to the other side before she could get me. 800 pounds of a mean porker isn't my idea of fun!

Now the panic began. Shana was inside the sow house with Rosie's babies. Rosie was inside the house but held back from Shana by the two by fours. She couldn't get in to save her babies like she'd wanted.

Shana was scrunched into a corner crying, thinking Rosie would eat her or something. Bill was opening the gate to let Rosie out so we could get her back inside her own side of the pen. I was in shock.

Bill did open the gate and let Rosie out. Her babies started piling out of the house with Shana right behind them. Shana was crying hard and looked like she'd been pulled through a ringer. I just wanted everyone to go home. This was too much!

But Bill wasn't done. The last male piglet ran by him and he grabbed it up and walked off with it. I didn't know what to do. That piglet was squealing like there was no tomorrow. Rosie was still outside her pen and now she went into a panic. She didn't like Bill and now he was holding a piglet in front of her, daring her to do something about it?

Well, Rosie didn't mess around. She ran up to Bill right away. Bill kept holding the piglet. I yelled for him to 'put that piglet down!' but he didn't - even as Rosie was running at him, grunting noises we'd never heard her grunt before.

Shana and I were screaming now, telling Rosie to stop and Bill to drop the piglet. No one was listening. Sheri just stood there watching her husband.

Rosie caught up to Bill and knocked him over with her nose! He fell backwards into a pile of compost and she came at him again! Bill was hitting at her; he had finally dropped the piglet but I don't think Rosie cared about the piglet by this time. Now she just wanted to teach him a lesson.

Shana and I were screaming for Rosie to stop. Panic had taken over. And my worst nightmare came true. Rosie bit Bill in the leg. She had him pinned on that compost pile and was yanking on him with her mouth! That sent me and Shana into more screams.

It's not like it sounds. There wasn't any blood on Bill, because Rosie didn't really want to hurt him, or she could have torn him to pieces.

After Rosie walked away and left Bill to his compost pile, things settled down. Bill had torn pants and bruises - that was all. Shana didn't want anything to do with Rosie -ever. I didn't know if I wanted Rosie on our homestead anymore after she'd bitten someone. Would she bite us too? How could we trust her now?

Sheri said that Rosie was just doing what any good mother would do -protecting her young. Rosie didn't hurt Bill like she could have, she just warned him with a bite. Sheri and Bill both said that Rosie could have torn him to pieces but she didn't because she wasn't a killer.

I suggest taking your males away from the sight of, and hearing distance of the sow. You won't know what she'll do until it happens and believe me, by then it's too late. I didn't think Rosie would hurt a soul. And maybe she wouldn't have if Bill wouldn't have aggravated her so much. All I do know is now I've lost the trust I had in my number one sow and will probably have to sell her.

I love Rosie. That same night I went out to check on her. I was afraid of course, thinking she would take one look at me and charge over the fence to eat me alive! But she didn't even come out of her house. Rosie was so stressed and hurt by what had happened that she lay in her house with her piglets and cried.

After entering the pen on the other side so I would be protected by the two by four boards between the sow and me, I dropped some lettuce over the railing to her and she did eat it but I could see the tears running from her eyes. Rosie sure was crying. My heart went out to her. I just wished today had never happened.

Pigs Eat Flesh and Bones

This is a short one, but you'll get the point of it. Back in the olden days it wasn't uncommon to see a sow run past your house with your kids' arm in her mouth! Could you imagine that? I can't. But it was very common. Farmers had to keep their sows in good fencing (which wasn't easy back then) and watch the sows with the kids. Pigs can eat human beings whole; skin, bones, clothes and all.

Also back then, sometimes when folks would get in a tiff the bullets would fly and they had to hide a body somewhere. Into the pigs' pen the body would go and there would be no evidence left over.

235

Grand-Daughter is Killed

When I told our neighbor that we were taking Rosie to the fair with her babies this year he got a startled look on his face.

"What?" he asked. "What for?"

He was wondering why we would go to so much trouble.

Back in 1957-1958, a switchman friend of the railroad raised hogs. Our neighbor lived down the road from them. The farmer had a four-year-old grand daughter who came to visit the hog farm. She helped her grandpa with the hogs every day. When there were babies she loved to watch them, constantly in awe.

There were old railroad boxes offered free one day and the man snatched up the housing for his pigs with a smile on his face! Free housing! This was a special feat back then, as the hog business a lot of times made the mortgage payment.

The farmer set out to cut the tops off the railroad boxes and make them into farrowing houses for his sows. The farmer cut the tops of the railroad boxes to a level of four feet high. This way sows or babies could not get excited and jump out.

Sometimes pigs can jump high if they are frightened. The farmer was pleased when he finally set up straw and put his bred sows in the train box cars. Soon the sows had litters of piglets.

Once the sows had their litters, the four-year-old grand daughter perched herself up on the box cars and watched the piglets play. She loved to see the babies.

One fatal night while everyone else was in the house finishing up dinner the little girl climbed up the box cars and slipped. No one heard her screams. She fell right in the pen containing two very mean sows and their litters.

The farmer thinks that one sow started biting, drew blood and then the others joined. Sows that are mean don't need to be provoked. These two sows completely demolished the little girl. By the time someone came out to find her, all they could find were bits of her clothing. The pigs had eaten her -bones and all.

When all was said and done the farmer took his rifle and killed every last pig he owned.

There should always be an in and an out to your pens. That doesn't mean you have to have two gates. It just means that you should at least have one. I don't know if the girl could have gotten out if she'd had a gate to leave by, she was pretty young.

Nanny Goat Takes Over

I don't know if the next story should be listed under humor or horror, it has a bit of both. I met these folks at Founders Day Rodeo grounds where they had set up to vend for the weekend. Wonderful people. Sharon Stepp and her husband were next to our set up at the Rodeo.

I set up as an author for the promotion towards _PIGS; and other stories_ and Sharon Stepp was more than happy to help.

Valentino took our breath away! She was a beautiful two-year-old who was being raised by her grandparents (the Stepp's), and they were doing an exceptional job, too, from what I viewed.

Sharon's sister, Lynn Minor, has had three milking goats for years. She doesn't need any more than the three because they keep her busy enough. (I can certainly relate to that)!

Well Lynn wasn't always in charge of her goats. Her husband brought home their first nanny goat and proceeded to teach Lynn the art of milking. The trick was easy enough and she learned quickly. After about four weeks her husband decided to let her take on the chore by herself.

Lynn went down to the milking barn and brought their nanny goat in to milk. But the goat had a different agenda for Lynn. The goat pushed Lynn up against the barn wall and held her there roughly. If Lynn moved, the goat pressed harder. Lynn began to scream.

Lynn's daughter, age sixteen, came running after she heard her mother screaming. She walked into the milking room, marched over to the nanny goat, grabbed its horns and then jerked it backwards so her mother could escape! The daughter yelled, "Move!" to the goat and she did, but the goat took her time about it.

It was clear to Lynn who was in charge. She needed to do something to prove to the nanny goat that she would be the boss from now on. Now that her husband wasn't going with her to the milk barn the goat thought she could push Lynn around.

After talking to her husband they both decided Lynn would need to wear black rubber boots that ran up the length of her legs. This way she could carry a stick in her boot and if the goat started to mess with her she would whack it on the hindquarters and show it who was boss.

Lynn wore her black rubber boots with a stick inside and didn't have any problems for weeks. Then one day she forgot the stick inside her boot. I guess goats are smarter than we give them credit for because that is the day the nanny decided to push Lynn's buttons again. Lynn was attempting to get the goat into her stanchion when the goat suddenly turned on her!

"No!" she yelled.

And instantly she reached for her stick. Even though it wasn't in her boot the goat backed off. Wow. Lynn was finally in charge! Good for you Lynn.

Old Pig Farmers

This one is a bit gross. I shiver to think it was like this back then. Bill Fisher used to help run the farrowing barn at a farm with 200 sows back in Shelbyville, Illinois. Mr. Fisher had this job for a few months until the farmer ran out of work and then he moved on down the road.

His first comment was, "You didn't drink the water that was for sure!"

"Why?" I asked.

And he told me the well was shallow and the hog pens were close enough that the water was contaminated by urine and feces. It was disgusting. Bill said the farmer would drag water from a pond nearby to drink. Now, that is sick, huh?

It gets worse. I was more concerned with how the animals, namely the hogs, were treated. Bill was in charge of the farrowing barn. He would help the hogs have their piglets and clean up after them.

Each morning the farmer would come out to view his herd of piglets. If there were runts, and usually there was at least one in each litter, the farmer would yank the helpless critter up by his heels and smack him against the cement wall until he died.

And as if that isn't enough... when a sow didn't have her litter on time the farmer would have Bill reach up inside her and yank out piglets in the birth canal. The mother was culled and babies were killed. Mother's who weren't fed properly would regularly eat their babies, too.

I was also informed of the twelve pens built for butcher hogs. These hogs were grown out to about 230 pounds and then slaughter. There were twelve pens, and each pen held 75-80 hogs. The farmer didn't want the hogs to move around at all because then they might lose weight.

One afternoon Bill and a coworker were sitting around goofin' off. Bill said they could hear someone screaming and went to see who it was.

It turned out some lady wasn't watching her three-year-old daughter and the girl was scared to death standing next to the sow hut where sows were coming over by the dozen to check her out. They were big and mean. Bill said if the girl had wandered into the pen the sows would have eaten her for sure!

What do you think about that? Got a scary farm story to share with me? Write and let me know.

Farmer Saved by his Gun

Another farmer knew his sows were nasty old hags. When the sows had litters of piglets, they were even nastier. So, he carried a gun in the back of his pants every time he had to enter their pens. He didn't believe in taking chances with his life.

Part of the fence was down in one of the pens and he had to go in to fix it. He knew the sows were mean when they had litters with them. He didn't want to fix that fence but he didn't want those mean beasts getting loose either! So he went inside to fix the broken fence.

Out of nowhere, and for no reason, a sow raced up and grabbed his arm! This farmer was older and not really strong. The sow had pulled him down into the mud and munched his arm a bit before he could reach behind him with his other hand and grab the gun.

He shot her, and the noise scared off the other sows which were starting to circle him. They all ran off

241

except the one he had killed. He butchered her up and had a feast with his neighbors!

19
Health on our Farm

If you take your animal home and it doesn't need to leave your farm again you may never have *any* health problems. Dirt and droppings from others' farms have different parasites which your animals may not be immune to and vise versa.

First Aid Medical Box

As always I need to write that I am not trying to tell anyone else how to run their own enterprise. When Ron and I started our agriculture business we started from scratch. Every book on the subject was either bought or borrowed from the library.

But the books seemed to have the basic information and the text was very complex for us. With *this* book I am hoping to give a little more detail and light on the subjects we have learned about. In no way do I think of us as experts. We learn new things every single day! There are a few things I strongly believe in.

- ♥ On every farm there should be a first aid medical box which contains the main necessities for health care emergencies.

- ♥ I keep only the minimum first-aid products needed because we don't seem to have many problems. However, since we have added to our herd of sows there have been more calls for medical attention.

I will list what I keep in our first aid kit and what we have come up against on our homestead. You can then ask your local veterinarian what would be best for you to have on hand at your homestead.

I use a tall plastic storage container with a handle on the lid to store my medical products in. I try to keep all the liquid products standing up, not laying down. This will help keep things clean inside my container. Here are a few things I keep on hand:

- ♥ **Wound-Kote Blue Lotion:** This is used when animals get scrapes, cuts or abrasions. Any wound. It is an antiseptic/germicidal and cleanses the area.

- ♥ **Iodine Wound Spray 2.44%:** This is used to dip umbilical cords to prevent naval infections. It can also help with superficial wounds.

244

- ♥ **Vet-Rap Bandaging Tape:** This would come in handy in covering a wound if it was necessary to keep the wound clean and dry. I haven't had to use any of this wrap yet.

- ♥ **Antibiotic called Penicillin G Procaine:** This is used anytime an animal has an infection or if they are hurt and you fear it will become infected. This can stop germs and infection from spreading. It has to be kept in a refrigerator and is an injectable.

- ♥ **Catron II:** We use this after castration, or if we have any open wounds to cleanse against infection.

- ♥ **Vitamin B Complex:** I keep this on hand to inject my sows with during the winter when they can not get to the vitamins in the dirt. Actually I inject the vitamin into a piece of bread and feed the sows the bread. This vitamin is safe to administer to a nursing sow. I give each sow a 5 cc injection (into a piece of bread that I feed to them) for three days if I see signs that it is needed.

- ♥ **You may never need** to give your sow's the vitamin B Complex mentioned above, or any extra vitamins. I am told if you feed your sow's good quality alfalfa hay every day she will get all the vitamins she needs.

- ♥ **Different Size Needles and Syringes:** I carry a variety of sizes. You may want to use different product which your local veterinarian may suggest for your farm needs. For piglets I use a 22 gauge with a 3/4 inch needle and a small syringe to administer ½-1 cc. A sow's skin is tougher so I use

18-20 gauges with a 1 ½ inch needle and a medium syringe to administer 3 cc's or more depending on the dosage of the product. *I never reuse needles.*

♥ **Scissors:** I keep a cheap pair of scissors on hand in case I need to cut away the excess skin around a wound. The scissors I use have a flat and a sharp edge.

For us this is enough to have on hand with the emergencies we have come up against thus far. You may want to prepare your first-aid medical box with a more extensive list of items.

Shana Learns to Castrate Piglets

My good friends, Bill and Sheri Vejrostek, have once again turned a terrible process into an excellent learning experience. I always look forward to visiting their farm.

Today Shana accompanied me to learn about castrating piglets. Ron and I are big babies when it comes to blood or anything remotely close.

However, I can't see paying the veterinarian $10.00 per piglet to castrate. I feel this will be a huge expense when we have six sows here! We have four breeding sow now and the males born could add up to twenty or more.

Shana agreed to learn how to castrate so she can do the job for Ron and I. Bill is going to teach her to castrate the piglets at three-weeks of age.

Sheri likes to wait until the piglets are five-weeks-old. The piglets are taken away from mama and readied for their weaning journey. Castration is the first part of their weaning process.

This time Bill talked Sheri into letting him perform the procedure at three-weeks of age to see if it is easier on him and the piglets.

Don't tell your husband/boyfriend or any male you know, but there is *no* anesthesia given! This is the one fact that keeps my hubby in the house!

♥ Bill insists on using all sanitized equipment. We needed a clean one gallon bucket with hot betadine water, sponge, scalpel and a towel.

♥ Shana and Bill castrated five males in less than half an hour. Bill brought one male piglet into their kitchen.

♥ Shana grabbed the piglet by its hind legs and swung it so the piglet was upside down, the head was placed under her butt and Shana could hold the back legs apart for Bill to see clearly. Shana sort of just sat on the piglet while holding its back legs firmly in place in front of her.

♥ They did the dirty job right in Bill's kitchen and it wasn't messy at all.

♥ With Shana holding the legs tightly in place, Bill squeezed the testicle so it was taught against the skin and made an incision next to it. This way after the incision was made Bill could push the testicle through the hole he created. There wasn't much blood if any. I was surprised.

♥ Then Bill shaves off the muscle from the cord until it disconnects from the body. This cord is a mucous-white color.

♥ Bill advised us not to be in a hurry at this point. You wouldn't want to cut the cord off without shaving and have the pig bleed to death internally.

♥ After shaving the cord slowly until it disconnects from the piglet, repeat the procedure on the remaining testicle. Then spray an antiseptic on the area and let the piglet return to his siblings.

Piglet Loses its Face

My beautiful sows cannot be careful enough when they have large litters. Puller has proven this to me with her first litter on our homestead. She is extremely careful when standing or lying down; but with *thirteen* little ones there isn't much room to move about and Puller stepped on a piglet!

♥ I know it must have hurt the poor little female when it happened but she didn't die. Puller had stepped right on her face. The skin was pulled downward and dragged loose into a pile next to her mouth.

I thought, "I have to help this piglet!"

I couldn't do anything until the next day because we were on our way out. I knew the vet wouldn't be able to help the piglet since the wound had healed over and there wasn't any sign of infection.

After we got home I called my trusty 'pig lady' friend Sheri Vejrostek and asked her about it.

"I wouldn't take her away from mama," Sheri said. *"Try and take her aside to clean up the sore and add antiseptic if necessary. Then put her back. She's getting the best care from mama though, so I wouldn't take her out of the litter. You could make her a bowl of milk with a little oatmeal in it for an extra boost."*

"Okay. Thanks!" I said. *"Hey. Sheri? How are your babies doing?"*

"Oh! They are doing just great! Mama isn't bothered when we pick them up. I take a dog brush and sit with her for hours. She will roll over on her side so I can brush her belly. It's wonderful! Good luck."

"Thanks." I said.

I would need a lot more than luck! That baby hadn't been eating because it was too painful for her. Like when I'm sick -I usually don't eat much. Same for this baby. Only I think it took a huge toll on this piglet. She was under a lot of stress.

Also, I decided to give the piglet another dose of iron. I made up a box full of cleaners and milk with oatmeal in a bowl, the iron dispenser and a wash cloth to clean up the babies face. I decided to clean it as best as I could and see what the damage was.

I asked Shana to come with me to help hold the baby while I did the cleaning. We got dressed in our usual winter attire; coat, hat, boots and gloves.

"She's really going to squeal, mom." Shana said.

"I know, hon. But we have to get her cleaned up if we expect her to live."

Within minutes we were at the pigs' winter housing. I could see Puller (mama) standing just inside the doorway. She was chewing on something.

"Puller's eating that dirt you brought in yesterday. Guess you'll have to get more." I said.

But it wasn't dirt Puller was chewing on.

"What is that?" Shana asked me.

"It looks like a piece of cardboard box." I said.

♥ We got closer and Puller came out of her pen to see if we'd brought her grain. Puller had a small spine hanging out of her mouth and she sort of slurped it up into her mouth like my kids do with spaghetti noodles!

Shana gave both the sows grain so I could look inside the house without any troubles. Who knew what to expect? I didn't know if she'd eaten the sick baby or what. I was completely mystified! The piglets were thirteen days old. Why would Puller kill her baby now?

♥ Usually if a baby is deformed or not healthy, the mother pig will take its life shortly after birth.

250

- ♥ It's not normal for a sow to wait thirteen days and then eat a baby!

- ♥ It turned out that Puller was doing her duty as a good mother in a bad situation.

Puller had stepped on the piglets' face and the wound became infected from the urine in the straw the piglet had laid upon. Puller knew her baby would not live but she didn't kill it. The baby died during the night and then Puller ate it. This is normal.

- ♥ If a baby dies some sows will consume its body.

- ♥ Sometimes a vet will recommend that you feed the dead baby to the mama to help build up an immunity towards whatever the baby died of.

When I looked inside the house doorway I saw the last of the piglet. It had been dead for a long time. Severed in half. Puller had eaten the bottom half of the piglet and I could see inside the body cavity. It was pretty gross, I will tell you that! I grabbed the front feet with my gloves and removed the piglet from the structure.

The other babies were in the corner under the heat lamp. I counted piglets to make sure there were still twelve, and there were.

I was so upset by this time that my throat was choking up with tears.

My advice to you? If you have an injured piglet, by all means, *clean it up!* You really have to let Mother Nature do her job with anything else, though. No one wants to see a baby die, especially a baby as cute as that piglet was.

You'll be glad to hear the rest of the bunch is healthy and happy as twelve peas in a pod!

Cauliflower Ear

If you have this problem you will know what I'm talking about. Otherwise, I hope this will explain it significantly. We have a piglet whose ear was hurt. It was either bitten by its litter mates or stepped on by its mama. Can't tell which one it is.

♥ What we *can* tell is that the ear is three times the thickness as it should be. The ear itself is the same size as the other ear but the *thickness* of the skin is over an inch thick. Its opening to the ear canal is nearly closed off. I have to wonder if the baby can hear at all.

♥ At any rate, I asked the vet about this. Cody Ames, D.V.M. of Tonasket Veterinary Service informed me that this happens to dogs if their ear is bitten or scratched open and then closes before healing completely.

♥ The inside of the ear within the two flaps of skin will then fill with fluid. The vet is then usually called. It's not an emergency although they lance the ear, drain it and put the dog on antibiotics.

Cody told me that pigs have an awful lot more veins in their ears than dogs do and so the procedure above would be a last resort. If I lance the ear and drain it the chances for infection attacking the entire litter are much higher, too. He suggested that I start by injecting the piglet with ½ cc of antibiotic each day for a week.

Cody's done an exceptional job and he is more than willing to explain things to his customers so they might cut costs by doing simple procedures for their own animals at home. I think Dr. Ames is great.

Mathew Deebach explains further...

Of course we followed Cody's prescription for the swollen ear.

♥ I injected the little piglet each day and watched as the swelling in the ear shrank.

♥ Shana would hold the piglet as still as she could and I injected it in the muscle area just behind its ear.

It went pretty smoothly considering I am deathly afraid of needles. I had all three of my children natural, no drugs and therefore no needles. It's getting a bit less frightening for me with all the shots I've had to give our piglets. I don't sweat anymore!

We found out some more about the 'swollen ear' condition, too. One of Shana's friends, Web, who lives near us, came to see our captivating attraction of twenty four piglets last week.

Web told Shana, "That pig has cauliflower ear."

Shana, nor I, knew what this meant at the time. Shana had not asked Web to elaborate on the subject.

253

We called our local agriculture teacher, Mr. Deebach who won the teacher of the year award in 2001! He was nice enough to fill in the gaps for us.

Mathew Deebach, Tonasket High School FFA teacher

Apparently it truly is called 'cauliflower ear'. The reason this happens originates from the piglet being smacked or bumped hard against his ear.

- ♥ There is a cut or scrape and the affected area closes.

- ♥ The ear's skin is separated from the cartilage and fills with fluid.

- ♥ It becomes bumpy, hard and thick looking.

- ♥ As long as no infection is present there is no treatment necessary for this type of ailment. The ear will lose its swelling as time goes by.

Abscess

With two sows on our homestead we have come across a few more health conditions. In one litter of piglets we had a male with an abscess on his ham (his back hip area). It was a sac of squishy fluid and the skin jiggled when the piglet walked about.

Not knowing what the abscess was I looked the rest of the piglet over from head to toe. He had a few things going on.

♥ First of all I had noticed the sack of fluid on his hind end.

♥ Second, there was a nasty cut on his cheek just under his eye.

Then, after spraying the gash with antiseptic and letting him go with an injection of Penicillin for two days, we checked his wounds again.

This time we discovered the sack of fluid was shrinking, the gash under his eye was a bit yellow in color and now the piglet's eye was swollen completely shut.

I called our veterinarian, Cody Ames, D.V.M. Cody never seems too busy to help us out and that's just one of the great things about him! Hopefully you have a good veterinarian in your area. A lot of times you can take care of mishaps yourself *with instruction.*

♥ Cody asked me to flush out the eye and the wound just below it. Clean it up as best as I could. Then put some of the Penicillin in the wound and into the eye.

♥ This would help clear up the infection faster since there seemed to be multiple problems.

♥ Cody then told me to increase the Penicillin injection and put the baby back with mama.

I would keep an eye on the wounds and hopefully the piglet would be able to open his eye soon.

Of course we did as we were told. The piglet didn't squeal while we flushed out the eye. There was a small amount of white puss in the eye area. I flushed the wound below the eye as well then applied a wet towel to remove the yellow colored mess. Our baby was the perfect patient. He didn't yell or squeal or even pee on us!

Shana with both piglets on antibiotics

Flushing and cleaning the wound below the eye was the easy part. Now I had to give the injection of Penicillin. I hate to hurt the piglets and it does hurt to give them an injection. Just like it hurts us human's a little. Our little guy did cry out a little at this point but we did it fast and he quieted down just as quickly.

♥ I squirted the last of the injection over the wound site and around the eye area. The piglet could open his eye about ¼ of the way by now.

♥ He was watching us as we worked on him. This piglet was a lover, too. He snuggled his nose into Shana's coat as he grunted his displeasure at us. But he hardly protested at all.

Shana and I took the piglet back to his mama and let him rest. This was a work out for such a little guy!

Injection for the Piglet

For a baby piglet you will want to consult your veterinarian before injecting medications or iron. Every piglet is different in size and strength.

The area where you live may carry different parasites which could cause your piglet to need vaccinations. We don't need to vaccinate where I live. I am no expert in this area, I am learning as the days pass by.

It is incredible what an animal the size of a five pound piglet can endure. They are amazing as I keep stating. Since the piglet (a few paragraphs above) was stepped on and Cody suggested we give the piglet an injection of antibiotic's each day for a week, I had to follow through with the order.

It is not an easy task to give an injection. The process is easy enough. But if you have a weak stomach for needles then the process gets a bit complicated. I have had practice this past year giving injections and it has become easier. Each time I am presented with the task I continue to get better.

♥ Shana assists me daily. She wants to be a veterinarian you know. We caught a piglet and brought it outside of the pen to assure mama didn't attack us if/when the piglet squealed.

♥ Piglets have a set of lungs on them that could break glass!

♥ Shana held the piglet close to her and I found the muscle behind the ear. The area is thick and soft, there are no bones felt underneath. This is the muscle area where you should inject the needle ½ an inch.

♥ Pull back on the syringe a tiny bit to be sure there is no blood. Blood would indicate you hit a vein and you need to pull the needle out and start again.

♥ If there is no blood when drawn back just a little on the syringe, then plunge the medicine in and withdraw the needle. You are done!

Crushed!

When you have a sow that farrows *fifteen* babies you are bound to have a few complications. For us this meant our sow Rosie would be stepping on her piglets.

She had far too many to keep tabs on. Rosie is an exceptional mother and she loves all of her babies.

The piglets are smaller, weaker and slower to move when there are fifteen of them. This is why Rosie could not keep track of *where* all of her babies were.

♥ Usually piglets are running and playing in the farrowing house by the age of one-hour.

♥ With smaller piglets they are not moving as quickly and can not move out of mama's way when she lies down, no matter how careful she is.

Rosie farrowed fifteen babies but one was born dead. She had fourteen alive when she was done farrowing. Sometime during the early morning Rosie laid on a piglet and it died. Shana found the baby next to Rosie's nose the next morning. Rosie was trying to nudge it back to life.

The following morning Rosie stepped on another piglet and this baby also met its maker. It looked like Rosie had laid on the piglet, then while moving quickly to get off the piglet she stepped on it again and squished some of its intestines out on the inside of one leg. She seemed extremely upset about this baby dying.

We are blessed with twelve beautiful healthy babies, since one died during birth. And we couldn't be happier.

♥ I do plan to move Rosie to a bigger farrowing pen for her future litters.

♥ My daughter and I will take turns making sure the babies are under the light and not able to get trapped under mama for the first few days.

♥ Once the piglets are a few days old they'll be quick enough and will have learned to run when mom comes near!

Choking Piglet

We had a four-day-old piglet come to live with us. Rayna was only a pound in weight at birth, possibly a bit more but not much. She wasn't doing well in the litter and we took her home to try and give her one-on-one care since the owners had nine others to deal with.

During her fourth day with us we nearly lost her. She had gotten something in her throat. We still don't know what it was.

♥ Rayna was frothing at the mouth. Her outer eyes, lips, snout and mouth area were a light blue color. I knew she wasn't getting any oxygen.

♥ I tried pushing on her belly hoping the object would shoot out of her mouth like in funny movies on TV. But that doesn't happen in real life!

♥ She was choking and I couldn't seem to help her. I almost started crying. So did the girls as we watched Rayna's mouth froth and her eyes were watching us -seeming to beg us to help her. It was heart wrenching.

♥ I thought about it quickly and decided the only thing to do was to stuff my finger down her throat. I did it. I jabbed my finger down her throat and pushed her tongue back out of the way.

♥ Right away you heard Rayna inhale deeply and kind of cough. She could breathe. Whatever I did it helped and she lived. A choking pig is no fun time!

Pneumonia

Piglets need a heat lamp during cold winter months. A slight chill is all it takes for an entire litter to get sick and they can easily die.

This past winter some friends' sow knocked over and unplugged the floor heater which had been set up. All ten piglets were in thirty degree weather over night until the farmer found the mishap the next morning. Seven piglets died of pneumonia.

Ed Maleng of Tonasket, Washington bought four piglets in the spring from us. It was still forty degrees at night and he only used straw to keep them warm. They were fine for a few weeks but soon caught a cold.

♥ Ed said he saw one pig sort of choking or seeming to try and cough up a hair ball.

♥ He left for the day to work with his buddy and when he returned the piglet was dead. Ed figured the pig choked on some food.

A few days later the vet had to come and administer a shot to another pig which had started the same coughing spasm. Ed didn't want to take the chance of all four of his piglets dying. He asked the vet, Cody Ames, D.V.M., what could be wrong with his pigs. Cody said that there is a disease that some litters get but it is rare and he didn't know if this was the case or not.

While talking with Ed at Maverick's Bar & Grill we finally came to the conclusion that there wasn't any heat for the piglets and it was still forty degrees in the early mornings. Ed admitted that he thought the piglets caught a cold.

♥ Piglets should be kept in temperatures above 50 degrees! They are not prone to cold nights.

♥ If you keep a ten inch bed of straw for them to bury into, this will add ten degrees to your outside temperatures.

Lice

Our first encounter of a problem with Rosie came when we hauled her to another farm for breeding. Rosie was eight-months-old and had been on our farm since she was six-weeks-old. We'd never had a health problem.

♥ I did check out the other farm before taking her there but the pig which we bred her to was white and lighter colors make it much harder to detect lice. He looked and acted healthy.

♥ I didn't look closely for lice bugs. I couldn't see the eggs on him since his hair and skin were white. He was indeed infested with lice and Rosie was with him for ten days.

♥ When Rosie came home I didn't think to check her for lice eggs.

It was a week later when I saw the eggs stuck behind her ears and down along her jowls. It's amazing what a tight grip those things have. I saw one bug when I investigated. I am thankful she had only one bug because that was enough to make me sick!

For two days I picked those gross eggs from behind her ears. It was terrible. I used my nails to scrape them away. It wasn't easy either. I also sprayed her entire body with Zonk It which you can purchase at a feed store.

That episode taught me a valuable lesson. From now on I will spray for lice before sending sows to another farm hoping that any bugs will be repelled from her. I will also spray her *immediately* upon coming home.

Injection for Sows

Rosie did not like the injections! I tried to trick her but it's hard to trick a smart sow.

♥ I got a number eighteen needle, filled the syringe with 5 mm of Vitamin B12 and attempted to give her the shot above her front leg and behind the ear in the muscle area. She did not like it one bit, though.

♥ I didn't press it in hard enough so she grunted her displeasure and walked away. I was left holding the syringe with vitamins dripping out and turning my fingers orange!

♥ You do have to apply pressure when giving a shot to pigs. Their skin is as tough as nails! Just hold the syringe so you are aiming it over its destination and push it in hard.

♥ Try using one hand too, so the second the syringe is in you can push the plunger in to release the medication with the free hand. You have to be fast or the pig will walk off and the medication will be lost. You may not get a second chance either.

I suggest putting a tad bit more vitamins/injectable in the syringe and do the event as quickly as possible. If she walks off some of the medication will be lost on her skin but the bulk of it will get to where it's supposed to go.

Trim Pig Hooves

I have yet to find a book that tells about trimming hooves on pigs. I find it hard to believe no one else has come across this problem.

♥ In big confinement pens the floors are concrete and this will wear down the hooves naturally.

♥ If your outside pen has rocks or hard ground for your sow to walk on she can wear down her hooves that way.

Rosie needs her hooves trimmed by me. I don't mind doing it either. It's simple. But it does take time *and* patience. I asked the vet to come up here the first time and show me how. Rosie was eighteen months old. It cost me an arm and a leg and the vet couldn't get a lot done while here. I got the idea of how to perform the task, though, and that's all I needed.

♥ First you will want to buy some sharp horse clippers at your local feed store. Mine cost $32.

♥ While your sow is sleeping on her side you take just the end of her hoof in the clipper and use the same method you do while trimming your own nails. Make sure you only take small amounts off at a time, stopping when the nail becomes lighter.

♥ The back nails are the most important. It is painful for pigs to put pressure on those back nails. If your sow limps or doesn't want to get up, you can try trimming her nails to see if this helps.

♥ Of course, exercise precaution while using this technique. Not all sows are friendly and if you cut her too short she may get upset just like you would! I take very small pieces off at a time. Sometimes it takes me three or four days to finish with all four of her feet. You will need patience when the pig wakes up and stands if you have not finished trimming.

♥ It will help if you make her pen wet for a day or two before you trim. The water will soak into her hooves and make the trimming process much quicker!

Mastitis

Mastitis is no laughing matter. Milking goats is a lot of work and to have your production slowed down or lost forever is too much. I was nearly in tears when our milker Katie came down with mastitis.

Luckily I had been reading every book I could get my hands on about goats. I noticed the symptoms before it got out of hand. Mastitis can destroy the milk producing tissue. The utter seemed a bit warmer to me for two days. Then the milk wouldn't just flow out as it had prior. I would try milking for up to five minutes and have a minimal stream flow. And her milk production was lessening each time I milked.

- ♥ I looked into the books and found that her symptoms matched mastitis and I got on the phone to get help immediately!

- ♥ The veterinarian in Oroville, Jacqueline Walimaki, D.V.M. advised me to: "Milk her out every 4-5 hours," because I caught the disease at the beginning of its cycle.

- ♥ Jacqueline advised: "Not to give any medications at this time. Just milk her out 90% and make sure to milk her softly. Be gentle."

- ♥ I did exactly as I was instructed and Katie seemed to get better by the hour. By day four it was obvious that Katie would recuperate 100%.

Other Mastitis Advice

When something happens on my homestead I can't just sit back and let it get worse. I dive in and try and 'nip it in the bud,' so to speak. So I called everyone I knew who was knowledgeable about goats.

First, I called my neighbor Shannon but she wasn't home so I hung up after leaving a message. Second I called the lady I had bought Katie (the infected goat) from a few weeks ago. Terry Dean was visiting in Spokane with her son and his girlfriend. She was given the message that I had called and would call me the next day.

I was frantic. I didn't think that milking her out every couple of hours would be enough. Everyone had told me before that mastitis was something to be concerned about and to be aggressive and take care of the problem.

Of course, Jacqueline had given the correct advice and things would clear up in time, but I didn't know that the first day.

Finally I called another goat owner, Jessica, who lives just outside of Tonasket. She was home and talked to me at great length.

♥ Jessica's opinion on mastitis was to give the goat a 6cc shot of Antibiotic's once a day for three days and milk her completely dry.

♥ Not being sure about giving an injection I hesitated. I didn't want to give Katie an injection in case I wasn't right to make this decision! Why is this always happening *to me?* But on the other hand, I didn't want to have an epidemic break out and have *all* my goats come down with mastitis.

We were losing milk production daily with Katie. I didn't want to loose the milk from the other two goats, too. Still I hesitated because I didn't know Jessica well and wanted to be sure about the injection. After all, the vet herself told me not to give any medications.

Milking out Katie every few hours was not good for my already bruised and swollen thumb and forefinger since I am a beginner in the milking area. Goat milkers use muscles that they never knew they had before a goat teat was placed in their palms. It's amazing how the fingers cringe at the very thought of the milking hour approaching.

My thumbs and forefingers would not work most of the time for four days after I started milking my three goats. I got cramps. My fingers tightened and wouldn't open up for a few minutes. My forefinger on the right hand was twice the size of the forefinger on the left. It was not a fun time. But I loved going out to milk!

At last Terry Dean was home and called me back. I didn't act on her advice because Katie was getting better by milking her out every few hours as the vet had prescribed. Here is Terry's advice:

- ♥ Terry advised me of what she did two years ago when she had mastitis in her goat barn.

- ♥ She gave each doe 6 cc's of LA-200 Antibiotic once a day for three days in a row.

- ♥ Terry also milked them out completely but *only* at regular milking times.

- ♥ The antibiotic will kill the disease causing the mastitis and the goat will get better quickly.

After second day I could see a difference in Katie. It was easier to milk her out. Her flow wasn't hard and pausing with the clumps hidden inside. She was finally clearing up. I was ecstatic!

By day three I was milking her out in fifteen minutes again, not the ½ hour it was taking me the past four days or so. Katie was better and neither of the other two goats became infected. Mastitis is contagious and will spread if not treated.

Other books by Jeanie Whiting

Pigs: and other stories
Farm Animals, Your Guide to Raising Livestock

Documents & Newsletter

See Resources

Jeanie Whiting was born and raised in a farm environment. She learned early on to respect and provide the best care for all animals. It has always been her belief that if you take care in the handling and management of your animals, you will be repaid three times over by the enjoyment the animals will bring you. Well-taken-care-of-animals are happy, healthy and a pleasure to be around.

Jeanie wants to dissipate a message with this book. She wants to tell the world that animals are worth loving, *even* if they are raised for consumption. If you take the time to love an animal while it is in your care, the animal will in return grow healthier and happier which will be the best return of your money spent.

For the past seven years Jeanie (along with her family) has raised pigs, cows, goats and milk goats, chickens, turkeys, ducks, rabbits and emu birds. They have two dogs to chase off deer, coyotes, cougar and bears away from their home in the mountains.

Jeanie provides a quarterly newsletter jam packed with farming ideas; hints and tips on raising animals, jokes, riddles, and contests through email for all subscribers. To subscribe for one year, send name, address, email and a check for $8.00 to Fox Mtn Publishing, PO Box 1516, Tonasket, WA 98855. Or email us for a FREE copy. When ordering, don't forget your relatives and friends. Great gift!

What is a day like on the Farm?
Here's an example! (And a FREE chapter from Jeanie's other farm book: *Pigs and other stories!*).

Every day is a crazy day!

'Just-A-Little Homestead' is the name of our home in the mountains. We will share our experiences with those who want to learn about pigs, other animals, children and family in the wilderness. Yes, we moved from the city and it has been quite a change for us. It is *always* an adventure!

Just the other day I was wondering why so many things happen to us at once. It's not like we get *one big change* a month around here, there are windfalls of big-happenings!

Like last Wednesday! Shana, Amorette and Kyra (my daughters) were helping me in the yard. We're getting ready for winter. Where we live it gets three-four feet deep in snow, and it stays for months. Here is an example of a normal day at our homestead!

7:00 Awake. Dress for the day. Clean up bedroom and come downstairs.
7:30 Amorette-feed animals, outside chores. Shana-cook breakfast.

8:00   Eat breakfast. Clean areas. Kyra has a bath after we eat.

8:30-11:30 Shana and Amorette do school curriculum. Mom and Kyra do chores.

1:30-12:00 Lunch. Clean up.

12:00-   Outside work - different each day.

Listed below are today's work accomplishments.

I finished splitting the firewood that my husband-Ron-had started last week before he left to work for the week. The kids and I are alone during the week. A neighbor boy had finished rototilling the garden for me. I still needed to plow the rest but I was hoping Ron would do that this weekend. I went to check on the other girls. Amorette was finished using the staple gun. She stapled plastic to the rabbitry to keep the wind out this winter. I told her what a great job she did.

Shana and Kyra had used a hand saw and cut down four dead trees for me. We drug them over and started a pile for burning when the snow is deep this winter. The four of us moved one and a half tons of hay from one stall in the barn to the other. We need to move the baby calves to the stall with a dirt floor for easier clean up.

By now, the girls and I were ready for a nap! It had been a busy day. But next week our new sow comes home and we need to build her a pen. So we couldn't take a nap. Shana and Amorette removed green t-posts from the ground where we didn't need them and we hauled them up to the building site. I was trying to figure out the best way to incorporate the pen into our yard, when our part-time neighbor came running up. Greg Short was at their cabin for hunting season this week.

"Is your pig supposed to be over in our field?"

"No. She's not!" I said.

I went running off with him to round Rosie up.

"We let her out so we could build a pen next to hers." I explained to Greg.

"I tried to turn her around and get her to come back this way but she's a lot bigger up close!" He said.

"Yeah. I know." I said.

I laughed. Rosie is our 500 pound sow. Rosie is very tame, as all of our animals are. But if you had a 500 pound pig running towards you, it *would* be intimidating! Like Greg, you'd move aside and let the pig pass.

When we got down to our gate I saw that it was broken. No wonder Rosie was out wandering the neighborhood! Our fourteen foot panel had come off the hinge and was hanging open at a crooked angle. Not a pretty sight where I live!

"Great." I said, just staring at the fence. I didn't have time to *fix* things, I needed to *build* things!

"I'll get a wrench and fix that after we get your pig home." Greg said.

"Thanks." I was really glad.

Rosie ran 600 feet through Greg's field, 200 more feet down a hill and across a very wet creek and another 50 feet towards the Cape LaBelle road. What a busy girl!

Have you ever tried to push a pig to go in a particular direction? It's amazing what strength they have to go the complete opposite way you intend! And that is what Rosie did today. She didn't listen to me when I told her I needed to get home to finish building that pen for her new room mate. All she cared about was the new earth under her feet! I thought of that song, 'I feel the earth move under my feet,' as she ran by me.

Shana, Amorette and Kyra came riding around to Cape LaBelle on our four-wheeler just as Greg and I were trying to get Rosie to head back home. She liked the creek and all the mud. It wasn't going to be easy. Why didn't I grab a bucket of grain before I left?

The sun was starting to come down now and I had Araia coming to buy our last horse in another hour. When was I going to get time to build the sow pen? I was pretty frantic by the time Rosie headed towards home again. My hair must have been standing on end!

Greg and I sounded like we'd run a marathon when we got Rosie back to our gate. She went right in to our property and made a be-line for the garden.

Then the problem came about fixing the gate. He'd stopped at his truck and grabbed four wrenches but they were all too big, he needed a smaller size. Back to his truck he went, another 1000 feet round trip. I don't think he caught his breath for a full ten minutes after the gate was back in place! I told him thank you for the help and we talked a bit.

By now it was time to do evening chores, and I had Araia coming to buy our horse, I still needed to build the pen, and I had to get Rosie back to her own pen! There aren't enough hours in the day around here!

Araia showed up right on time. She saddled our horse up and took her for a ride. I needed to be close by in case there was a problem, so I hung around the barn.

Rosie was still munching what's left of our garden. We have a 70' x 150' garden space.

Shana asked me to do the barn chores since I was at the barn. She went up to the chicken coop and rabbitry to do chores. Amorette took care of feeding Rosie, even though she was still in the garden. I picked up the pitch fork and started cleaning out the mess the calves left us since morning.

I gave the calves a clean stall, fresh water and two buckets full of grain. I added clean straw to their bed. I ran the hose to the outside water trough for our other cow, Bailey, to drink from and tossed Bailey a couple leafs of hay for supper.

"I love her!" Araia sings to me as she rides up on the horse. She really was singing; I'm not exaggerating! "When can I take her home?"

"Right now!" I say, thinking that it would be nice to have one less animal to worry about right this minute. I turned off the water and wound up the hose.

Shana came down with the phone in her hand.

"Mom! Ron is on the phone."

"Okay, hon. Just a ..."

All of a sudden we heard this loud noise. At first I couldn't tell what it was. Then I realized it was the chickens! I had never heard them make all that fuss before. They were cackling so loud it reminded me of a daycare center full of screaming two year olds.

Shana tossed me the phone and ran like the dickens up the hill. I would have a heart attack if I ran up *that* hill! It's the steepest hill we own! I stayed where I was.

"Ron?" I said, "I have to call you back. All hell just broke loose!" I hung up and turned to put the phone down on a bale of straw when Shana yelled.

"Oh my God!" She said. "There's a hawk biting our chickens' neck! It won't let go!"

All we could hear was the screaming of the chickens. I thought I could see feathers flying from down the hill where I was, but it was probably my imagination.

"Pepper! No!" Shana said.

"What's going on?" I yelled.

I didn't think I could survive walking fast, let alone running again after chasing Rosie a while ago. I would bet money that Greg was in a chair kickin' back over at his cabin.

I waited at the barn and listened. By now Araia had galloped the horse up the hill, past the house and down the driveway to our chicken coop. All I could hear was the steady pounding of horse hooves as Araia rode up. Things seemed to be quieting down for the night. Ha-ha.

Our dog Pepper went to investigate. He'd sniffed at the hawk and scared it. It let go of the chicken's neck and flew away. What a day. I went up to the house to get the papers for the horse sale and the phone was ringing. I picked it up thinking, what now?

"Jeanie?" It was Elanore Johnson from down the road. "I wanted to warn you about the two cougars we just spotted in the Mitchells' yard. Big ones."

That's all we needed. Cougars. Two cougars in a pair are very dangerous. Instead of taking a small calf, they will take away an entire cow. And we only *have* one full-time cow!

I thanked my lucky stars that Greg was nice enough to fix the gate for me. I reminded myself to get the horse papers and when I got down to the barn I would close up all the doors and lock the cattle in for the night. Now I had much bigger worries than building a silly pen for our new sow!

By the time I went to bed I was making a list for the *next* day. Bake a cake for Kyra's third birthday party. Wrap her presents. Decorate the kitchen for her party. Run to the store and buy ice cream. Spring clean her toy box to make room for all the new stuff. And then, if we have time *we still need a sow pen built!*

This is how every day is at our homestead. It is always busy and fast paced. You don't have time to wonder what it would be like to live in the city and sit in a nice recliner chair to watch TV at night. We don't get TV where we live. We have a VCR and we rent movies when time allows such a luxury. What the heck would we get done around here if we had to make time to watch TV every day?

Recipes I want to share with my readers

If you have a wonderful recipe you want to share, please send it to me. My address is in the Introduction page of this book. I may choose to add it to my next book!! Include your name, please, so you get the credit. Include any history that might come along with the recipe; i.e. if your grandpa stole it from an Indian during a coach raid and took it home to your grandma, who has passed the recipe down for decades!

Corn Bread
1 cup cornmeal
¾ cup whole wheat flour
2 Tbsp baking powder
1 egg, well beaten
1 cup milk
1/3 cup honey
3 Tbsp vegetable oil

Preheat oven to 425 F. Mix cornmeal, whole wheat flour, and baking powder together. Add egg, milk, honey and oil. Stir just enough to mix.

Bake in 2 greased 8-inch round pans for 15 minutes.

Chicken Enchilada's with special sauce

History.　　I used to live in a large neighborhood in Kent, Washington. I was starting over fresh from a violent relationship. I wanted my kids to know there was good in the world. I met a lady up the road from us who had two little boys close to my girls' ages. Mary and I started packing up the four kids daily and taking them to the lake, soon we were all good friends. I tasted these enchilada's one day at her house and I have loved them every since. Mary later introduced me to Ron!

1 pound chicken breast
1 cup mushrooms
¼ cup onion
2 cups cheddar cheese
1 pint sour cream
2- 10 ¾ oz cans cream of chicken soup
1- 4 oz can diced (mild) green chilies
Tortillas

Cut chicken breasts into bite size pieces and place in a greased frying pan. Cook 3-4 minutes - until it turns white. Sauté mushrooms and onion. Grate cheese.

Combine sour cream and soup in a bowl-set aside. Add green chilies to mushrooms and onion, and then add to chicken breast.

Combine all ingredients together into tortillas (except sour cream and soup mixture), adding cheese to each tortilla as you go.

Put into 9x13 pan and cover with sour cream and soup mixture.

Cook at 350 degrees for 25-30 minutes or until hot all the way through. Enjoy.

Want to make Your Breads Moister? Try these hints...

✓ Don't let your bread rise longer than necessary; too long a rising period can dry out the dough.

✓ Use potato water in place of any water called for.

✓ Add ½ cup mashed potatoes to your recipe.

✓ Use cottage cheese, yogurt or buttermilk as the liquid in recipes.

✓ Watch the baking time; don't over bake.

✓ Add peanut butter to your recipe.

Banana Nut Bread from Grandma Gray

History: Grandma Gray was a wonderful person. And she could bake. Every time we visited grandma she had goodies for us to eat. This bread was my favorite. It stays fresh in the freezer for months if I cook lots of loaves ahead of time and it's never dry!

3 ripe banana's
2 eggs, beaten
2 cups flour
¾ cup sugar
1 teaspoon salt
1 teaspoon baking soda
¼ cup chocolate chips, chopped
¼ cup pecan nuts, chopped
dash of lemon juice

Preheat oven to 350 degrees.

Grease bread pan. Mix banana and eggs together until smooth. Stir in other ingredients. Fold in nuts. Pour into pan.

Bake for one hour. (50 minutes if you make two small loaves)

Crab Cheese Dip

History: My Aunt Angie and I used to grub down on this dip all the time. Especially if we were upset. It just tastes so good. You can eat it on crackers. We used plain old saltines!

1- 8oz package cream cheese spread
2 tablespoons milk
Worcestershire sauce
6 ½ oz can crab meat, drained and flaked
1 green onion (scallion)

Combine: Cream cheese spread, milk and several drops of Worcestershire sauce in a 1-quart microwave safe bowl. Cook on 100% heat for 1-3 minutes or until cheese melts. Watch carefully.

Stir in crab meat and green onion. Cook 1 more minute, stirring frequently till mixture is heated through. Enjoy!

Makes 1 cup.

Cheesy Broccoli Tuna Bake

History: I got this recipe out of a magazine years and years ago. I don't remember which magazine, though, so if it looks familiar and you have the magazine, let me know! It's an easy, wonderful tasting tuna bake! One of the only ways I can get my family to eat tuna fish.

1- 12oz package Reames frozen egg noodles
1- 11oz can cheddar cheese soup
1- 5oz can evaporated milk
1   teaspoon instant minced onion
1- 12oz can tuna, drained
1- 4oz can mushroom pieces, drained (I use fresh)
1- 10oz package chopped broccoli, thawed
1   cup shredded mild cheddar cheese (I use 2 cups)

Preheat oven to 350 degrees.

Cook noodles in boiling water 20 minutes. Blend soup and milk into a smooth sauce. Add onions and mushrooms. Drain noodles. Layer 1 ½ cups noodles, 1/3 cup tuna and 1 cup broccoli in a 2-quart casserole dish. Pour in 1 cup of sauce. Repeat layers.

Bake 20 minutes. Then add cheese and bake 15 minutes more. Enjoy!

Makes 6 servings.

Homemade Kahlua

History: I received this recipe from a neighbor years ago. I make Kahlua every year to give for gifts to friends and family at Christmas time. And I try and make sure it is available for our Annual Round Up festivities. This Kahlua is used to make an excellent Kahlua cake, which is edible for kids and adults alike! (the recipe follows this one).

Boil: 3 ½ cups sugar
2 cups water   until it becomes clear, remove from heat and cool.

In ½ gallon container put a fifth of cheap vodka 80 or 100 proof, your choice.

Dissolve ¾ cup instant coffee (Sanka or Yuban) in small amount of hot boiling water (not in sugar and water mixture), add to vodka.

Put 5 tablespoons of vanilla extract or 2 broken vanilla beans into vodka.

Add sugar and water mixture to vodka, make sure it has cooled down slightly.

If it is made with vanilla extract it will be ready in a couple days; If made with vanilla beans it must sit 3-4 weeks. Shake bottle at lease once a day. Taste to tell when ready.

Kahlua Cake

History: Ron's dad recommended a man to build our barn. Since Pete came so highly praised, we let him stay in our RV trailer on our property. This saved him the 45 minute drive to and from town daily during his week of building our barn!

One night Pete invited us down to the RV to eat dinner with him. He had fresh beef for hamburgers. (This is the biggest reason I wanted to start prospecting with cows!) Anyway, his wife had sent him dessert. Guess what it was? Yes! Kahlua cake! She sent me the recipe right away and I've been baking it for my family at least four times a year!

Preheat oven to 350 degrees.

1 box yellow cake mix
1 large package instant chocolate pudding mix
4 eggs
1 cup oil
¾ cup water
½ cup kahlua
2 tablespoons sugar
1 tablespoon cinnamon

Blend cake mix and pudding mix together in a bowl. Add the rest of ingredients and beat well.

Grease bunt pan (I spray it with Pam). Sprinkle sugar
and cinnamon mixture into greased pan.
Add cake mixture to pan.

Bake at 350 degrees for one hour. Let stand 25
minutes, and then turn out onto a plate.

Have you ever tried....

Adding horse radish sauce and green peppers to your
meat loaf?

Joan's Broccoli Salad - Family Favorite

History: My mother's best friend, Joan Giancoli from Federal Way, Washington brought this recipe to my attention at her daughter's baby shower one year. It is the best tasting salad I have ever encountered. I made this salad for our own wedding in 1996 and everyone who tasted it from mine or Ron's side of the family wanted the recipe. If you don't like miracle whip, we've been told you can substitute with your favorite brand without noticeable results.

4-5 cups broccoli flowerette's
½ red onion - sliced
½ cup cheese - grated
½-1 pound bacon - fried and chopped

Dressing:
1 cup miracle whip
2 tablespoons vinegar
6 tablespoons sugar

Break broccoli into flowerette's. Slice onion. Grate cheese. Fry and break bacon into bite size pieces. Mix all the above into a bowl together. Mix the dressing in a separate bowl. Add the dressing mixture to the broccoli mix and place in refrigerator for at least one hour. Enjoy!

Oatmeal Cake

History: This recipe is from Esther Krohn of Chelan, Washington. She is a woman of many talents. When she gave this recipe to my mom and mom baked it for us, we nearly died from the watering our mouths made! This is excellent cake.

2 cups boiling water
1 ½ cups oatmeal (old fashioned)
Blend together and let stand

¾ cup butter
1 ½ cups white sugar
3 eggs
2 ¼ cups flour
1 teaspoon salt
1 ½ teaspoon soda
1 ½ teaspoon cinnamon

Cream together sugar and butter. Beat in eggs. Stir in oatmeal with dry ingredients.

Place in 9x13 greased pan. Bake at 350 degrees for 35-40 minutes.

Topping:
8 tablespoons melted butter
1 cup brown sugar
¾ cup chopped nuts
1 cup flaked coconut

¼ cup milk
1 teaspoon vanilla

Combine and spread on completely cooled cake. Place under broiler no more than 2-3 minutes. Enjoy!

RESOURCES

This is a short list of books, catalogs and other references which I have quoted in this book or I feel you would benefit from reading. I hope this can help you further research your subject.

Again, thank you for reading my book and please feel free to write your suggestions and stories to me at the address in the beginning of this book.

BOOKS

You can order any of the following books from Storey Publishing through the following address:

Storey Publishing Books
Storey Communications, Inc.
RR1 Box 105
Schoolhouse Road
Pownal, Vermont 05261
http://www.storeybooks.com

Storey's Guide to Raising Pigs $18.95
Storey's Guide to Raising Rabbits $18.95
Storey's Guide to Raising Poultry $18.95
Storey's Guide to Raising Dairy Goats $18.95
Storey's Guide to Raising Beef Cattle $18.95
Keeping Livestock Healthy $19.95

OTHER BOOKS:

Pigs; and other stories 158 pages, $9.95
Farm Animals, Your Guide to Raising Livestock 320 pages, $18.95

Send amount of book plus $4.00 shipping to:

Fox Mtn Publishing
PO Box 1516
Tonasket, WA 98855
www.foxmtnpublishing.com

Backyard Livestock $15.00
by Steven Thomas & George Looby
Countryman Press
PO Box 748
Woodstock, Vermont 05091

The Pack Goat $16.95
by John Mionczynski
Pruett Publishing Company
7464 Arapahoe Road, Suite A9
Boulder, CO 80303

CATALOGS

Jeffers   800-Jeffers
Hoeggar Supply   800-221-4628

NEWSLETTERS

Just-A-Little Ranch Quarterly   509-486-4919
$8.00 year

Fox Mtn Publishing
PO Box 1516
Tonasket, WA 98855

PERIODICALS

Mother Earth News Magazine
Ogden Publications
1503 SW 42nd Street
Topeka, KS 66609
(303) 682-2438

Countryside Magazine
W11564 Highway 64
Withee, WI 54498
(715) 785-7979

BackHome Magazine
PO Box 70
Hendersonville, NC 28793
(800) 992-2546

Small Farm Today Magazine
3903 West Ridge Trail Road
Clark, MO 65243
(800) 633-2535
Fax: (573) 687-3148

Today's Farmer Magazine
201 Ray Young Drive
Columbia, MO 65201
(573) 876-5205

HATCHERIES

Murry McMurry Hatchery
PO Box 458
Webster City, IA 50595
http://www.mcmurrayhatchery.com
Catalog: Free

Stromberg's Chicks & Gamebirds
PO Box 400
Pine River, MN 56474
Catalog: Free

WEB SITES

The vast web still amazes us all. If you want to use the web to obtain more information you will be surprised at just how much you can find. Here are a few websites that are available. Use your search engines to look up 'swine' and you will have all the data you need to keep you busy for months.

Homesteading and Small Farm Resource
http://www.homestead.org

US Department of Agriculture (USDA)
http://www.usda.gov

Cooperative Extension Services
http://www.reeusda.gov

American Livestock Breeds Conservancy
http://www.web.css.orst.edu

DOCUMENTS

Document #200 Farrowing List
A count down from breeding to farrowing day, including flushing the sow, using caution for 14 days after breeding, second belly stage, and lots more!
4 pages with resources $4.95

Document #225 Piglet Care
What to do with piglets after farrowing is over, giving iron, docking tails, clipping wolf teeth, castration, etc.
4 pages with resources $4.95

Document #250 Bringing Your Piglet Home
What to do when buying a new piglet, adjustment at a new home, housing, feeding, etc.   4 pages $4.95

Document #275 Small Farm Records

Records to keep for the small farm including Breeding, Health, Milking, Family Tree chart, Income and Expense, and more. You can copy these sheets, and get started today!   17 pages $8.95

For all documents above please send price plus $3.00 shipping to Fox Mtn Publishing, Orders Dept., PO Box 1516, Tonasket, WA 98855.

Cooperative Extension Service

For a lot more information on livestock and a list of their of programs in your area, write or call the Cooperative Extension Service in your state. You'll be amazed at how much information they have to offer!

Alabama
Auburn University
109D Duncan Hall
Auburn, AL 36849
(334) 844-4444

Alaska
University of Alaska
PO Box 756180
Fairbanks, AK 99775
(907) 474-7246

Arizona
University of Arizona
Forbes Bldg, Rm 301
Tucson, AZ 85721
(520) 621-7209

Arkansas
University of Arkansas
2404 N. University Ave
Little Rock, AR 72207
(501) 686-2540

California
University of Cal., Davis
300 Lakeside Dr, 6th Fl.
Oakland, CA 94612
(510) 987-0060

Colorado
Colorado State University
1 Admin. Building
Fort Collins, CO 80523
(970) 491-6281

Connecticut
University of Connecticut
1376 Storrs Road, U-36
Storrs, CT 06269
(860) 486-6271

Delaware
University of Delaware
Coop. Extension Service
127 Townsend Hall
Newark, DE 19717
(302) 831-2501

Florida
University of Florida
Extension Ser. Dept.
1038 McCarty Hall D
Gainsville, FL 32610
(352) 392-1761

Georgia
University of Georgia
Room 101, Conner Hall
Athens, GA 30602
(706) 542-3924

Hawaii
University of Hawaii
3050 Maile Way, Rm 202
Honolulu, HI 96822
(808) 956-8234

Idaho
University of Idaho
Coop. Extension Service
PO Box 442331
Moscow, ID 83844
(208) 885-6681

Illinois
University of Illinois
123 Mumford Hall
1301 W. Gregory Dr.
Urbana, IL 61801
(217) 333-5900

Indiana
Purdue University
Coop. Extension Service
1140 Ag. Admin. Bldg
West Lafayette, IN 47907
(765) 494-8489

Iowa
Iowa State University
315 Beardshear
Ames, IA 50011
(515) 294-6192

Kansas
Kansas State University
Animal Science
123 Umberger Hall
Manhattan, KS 66506

Kentucky
LivestockDiseaseCenter
1429 Newton Pike
Lexington, KY 40511
(606) 253-0571

Louisiana
Louisiana State University
Coop. Extension Service
PO Box 25100
Batan Rouge, LA 70894
(504) 388-4141

Maine
University of Maine
5741 Libby Hall Rm. 102
Orono, ME 04469
(207) 581-2811

Maryland
University of Maryland
1104 Symons Hall
College Park, MD 20742
(301) 405-2072

Massachusetts
University of Massach.
Extension Service
212 Stockbridge Hall
Amherst, MA 01003
(413) 545-6555

Michigan
Michigan State University
Coop. Extension Service
108 Agriculture Hall
East Lansing, MI 48824
(517) 355-2308

Minnesota
University of Minnesota
Extension Service
Room 240 Coffey Hall
St. Paul, MN 55108
(612) 624-2703

Mississippi
Mississippi State Univer.
Extension Service
PO Box 9601
Mississippi St, MS 39762
(662) 325-3036

Missouri
University of Missouri
309 University Hall
Columbia, MO 65211
(573) 882-7754

Montana
Montana State University
204A Culbertson Hall
Bozeman, MT 59717
(406) 994-6647

Nebraska
University of Nebraska
211 Agriculture Hall
Lincoln, NE 68583
(402) 472-2966

Nevada
Nevada Coop. Extension
Mail Stop 404
Reno, NV 89557
(775) 784-7070

New Hampshire
University of N.H.
59 College Road
103A Taylor Hall
Durham, NH 03824
(603) 862-1520

New Jersey
Rutgers State University
88 Lippman Drive
New Brunswick, NJ 08901
(732) 932-9306

New Mexico
New Mexico State Univ.
Coop. Extension Ser.
PO Box 30003 Dept. 3AE
Las Cruces, NM 88003
(505) 646-3016

New York
Cornell University
Coop. Extension Service
276 Roberts Hall
Ithaca, NY 14853
(607) 255-2237

North Carolina
North Carolina State Un.
Coop. Extension Service
PO Box 7602
Raleigh, NC 27695
(919) 515 2811

North Dakota
North Dakota State Univ.
Coop. Extension Service
PO Box 5437
Fargo, ND 58105
(701) 231-8944

Ohio
Ohio State University
Extension Ser. Dept.
2120 Fyffe Road
Columbus, OH 43210
(614) 292-4067

Oklahoma
Oklahoma St. University
Coop. Extension Service
139 Agriculture Hall
Stillwater, OK 74078
(405) 744-5398

Oregon
Oregon State University
Extension Department
101 Ballard Ext. Hall
Corvallis, OR 97331
(541) 737-2713

Puerto Rico
Univ. of Puerto Rico
PO Box 9030
Mayaguez, PR 00681
(787) 833-3486

South Carolina
Clemson University
Coop. Extension Ser.
103 Barre Hall
Clemson, SC 29634
(864) 656-3382

Tennessee
University of Tennessee
PO Box 1071
Knoxville, TN 37901
(865) 974-7114

Utah
Utah State University
Coop. Extension Service
4900 Old Main Hill
Logan, UT 84322
(435) 797-2201

Pennsylvania
Pennsylvania State Univ.
Agri. Admin. Bld.217
University Park, PA 16802
(814) 863-3438

Rhode Island
University of Rhode Island
Coop. Extension Service
12 Woodward Hall
Kingston, RI 02881

South Dakota
South Dakota State Univ.
PO Box 2207D
Agriculture Hall, Rm 154
PO Box 2207
Brookings, SD 57007
(605) 688-4792

Texas
Texas A & M University
Admin. Bld, Room 113
College Station, TX 77843
(409) 845-7800

Vermont
University of Vermont
Coop. Extension Service
601 Main Street
Burlington, VT 05401
(802) 656-2980

Virginia
Virginia State Univer.
Extension Service
101 Hutcheson Hall
Blacksburg, VA 24061
(540) 231-5299

Washington
Washington St. University
411 Hulbert Hall
Pullman, WA 99164
(509) 335-2933

West Virginia
Univ. of W. Virginia
PO Box 6031
Morgantown, WV 26506
(304) 293-5691

Wisconsin
University of Wisconsin
432 N. Lake Street
Room 601 Extension Bldg.
Madison, WI 53706

Wyoming
University of Wyoming
Coop. Extension Service
Room 102, PO Box 3354
Laramie, WY 82071
(307) 766-5124

Glossary

After Birth: The placenta and associated membranes expelled from the uterus.

Barn Records: A tally of daily, weekly or monthly notes in order to keep track of data.

Barrow: Male swine castrated when young

Boar: Uncastrated male swine.

Buck: A male goat.

Butcher: Refers to a hog being readied for, or sold on the slaughter market.

Castrate: To remove the testicles of a male animal.

Clean Pigs: Females and hogs up to seven months of age.

Colostrum: The first milk after farrowing, through which the sow/goat/cow is able to impart some of her natural immunities to the nursing pigs.

Conformation: Shape and structure of an animals' body.

Creep Feed: The early feeds, high in sugar and milk proteins that are offered to a pig while it is still nursing.

Cull: The removal of animals from a herd, usually due to poor performance or illness.

Dam: Another word for mother, used mainly with four-footed creatures.

Doe: A female goat.

Drake: A male duck.

Drift: A group of domesticated pigs.

Duck: A female duck.

Duckling: A young duck of either sex.

Elastrator: A tool used in castrating and docking. A tight rubber band is applied to the tail or the scrotum; the circulation is thereby cut off and the tail or scrotum gradually dries up and falls off.

Estrus: The period when the female will accept the male and conceive.

Farrow: The action of a sow giving birth to a litter of piglets.

Feeder pig: Young swine after weaning and before reaching slaughter weight. Between thirty and ninety pounds.

Finish: Final fattening of a pig up to the 250-260 pound market weight.

Flushing: The practice of increasing the feed intake of a female animal just prior to ovulation and breeding.

Free Choice: Free to eat at all times.

Gestation: The period of pregnancy, between 110 and 114 days.

Gilt: Young female pig.

Hay: Dried forage.

Heifer: Female cow that has not had a calf.

Hen: A female chicken.

Hernia: Also called a rupture, a portion of an organ or body structure has broken through the wall that normally contains it.

Hog: A swine over 120 pounds in weight.

Inbreeding: The mating together of closely related individuals to concentrate on certain traits.

Kid: A goat under one year of age; to give birth.

Litter: Piglets from a farrowing.

Livestock: Farm animals.

Market hog: The same as a butcher hog.

Mastitis: Inflammation of the udder, usually caused by an infection.

Needle teeth: Also called wolf teeth. These are two large teeth on each side of the upper jaw that are present at birth.

Pig Heaven: Dreamland for the greedy.

Piglet: Baby pig.

Polled: Naturally hornless.

Pullet: A young hen less than 1 year old.

Rooster: A male chicken.

Rumen: First large compartment of an animal's stomach, where cellulose is broken down.

Runt: Smallest piglet found in a litter.

Service: To mate animals.

Shoat: A term used for a hog from weaning to 120 pounds.

Sow: A female that has borne young.

Stanchion: A device for restraining a goat by the neck for milking.

Steer: Is a male cow that has been castrated before maturity.

Stress: A strain or tension on the animal's well-being caused by weather, moving, weaning, changes in ration, castration, or other environmental or physical changes.

Swine: A generic term equal to hog.

Udder: Encased group of mammary glands provided with a teat or nipple.

Vulva: The external part of the female genitals.

Weaning: To take piglets from their mother.

Wether: A castrated buck.

Wolf teeth: Same as needle teeth.

Yearling: Is a calf between one and two years of age.